Peace and War

PEACE & WAR

*Reminiscences of a Life on the
Frontiers of Science*

Robert Serber
with Robert P. Crease

Columbia University Press
NEW YORK

COLUMBIA UNIVERSITY PRESS

Publishers Since 1893

New York   Chichester, West Sussex

Copyright © 1998 by Columbia University Press

Library of Congress Cataloging-in-Publication Data

Serber, Robert

Peace and War : reminiscences of a life on the frontiers of science / Robert Serber
with Robert P. Crease.

p.   cm.

Includes bibliographical references.

ISBN 0–231–10546–0 (cloth : alk. paper)

1. Serber, Robert 2. Crease, Robert P. 3. Atomic bomb—United States—History.
4. Physicists—United States—Biography.

I. Crease, Robert P. II. Title.

QC16.S46A3 1998

[B]530'.092—dc21

97–38065

Casebound editions of Columbia University Press books
are printed on permanent and durable acid-free paper.

Printed in the United States of America

c 10 9 8 7 6 5 4 3 2 1

The George B. Pegram Lecture Series

To provide an opportunity for distinguished scholars to examine the interaction between science and other aspects of our culture and society, the Trustees of Associated Universities, Inc., established the George B. Pegram Lecture Series at Brookhaven National Laboratory.

The lectures are named in hour of George Braxton Pegram, who contributed so much to our country in general, and to Brookhaven National Laboratory in particular. Except for a few years abroad, George Pegram's entire professional career was spent at Columbia University, where he was a Professor of Physics, Dean, and Vice President. In 1946 he headed the Initiatory University Group, which proposed that a regional center for research in the nuclear sciences be established in the New York area. Thus, he played a key role in the founding of Brookhaven and became one of the incorporating trustees of Associated Universities, Inc., remaining an active trustee for ten years.

George Pegram devoted his life to physics, teaching, and the conviction that the results of science can be made to serve the needs and hopes of humankind.

George B. Pegram Lectures

1949 Lee Alvin DuBridge

1960 René Jules Dubos

1961 Charles Alfred Coulson

1962 Derek J. deSolla Price

1963 J. Robert Oppenheimer

1964 Barbara Ward

1965 Richard Hofstadter

1966 Louis S. B. Leakey

1968 André Maurois

1969 Roger Revelle

1970 Barbara W. Tuchman

1971 George E. Reedy

1972 Colin Low

1975 Jean Mayer

1979 Sir Peter Medawar

1985 David Baltimore

1988 Robert C. Gallo

1989 Sir Denys Wilkinson,
    Michael S. Brown

1990 Roald Hoffmann

1992 Maurice Goldhaber

1993 James D. Watson

1994 Robert Serber

Other Pegram Lectures Published by Columbia University Press

Contents

Introduction

Robert P. Crease

Robert Serber was a key partici-
pant in U.S. physics during a period that may well go down in history as its most productive period ever.

His upbringing was a Philadelphia story of a different stripe. He
grew up among a group of accomplished Jewish artists and intellectuals whose spiritual home was at 322 South 16th Street, the headquarters of "the Leof clan," as it was called: Dr. Morris V. Loef (a general practitioner with Socialist leanings), his common-law wife Jenny
Chalfin, and their three children: Madelin, Milton, and Charlotte
(Serber's future wife). Their house—generically a four-story brownstone, but the Leof's had it refaced in white stone—was a haven for
left-leaning artists and intellectuals in the 1920s and 1930s. The house
was practically a cultural force in its own right, the site of the first
"integrated" party in Philadelphia, play readings by Clifford Odets,
and conversation by the likes of musician Marc Blitzstein, author
Harry Kurnitz, poet Jean Roisman, physicist Wolfgang Pauli, and
journalist I. F. Stone. "Whatever happened in the world was immediately reflected in that house," Madelin Leof once said. Amusing
portraits of life at "322" can be found in biographies of several of its
patrons.[1]

Serber published his first paper in 1932, known as the *annus
mirabilis* or "miracle year" among historians of science for the number of events that occurred of far-reaching significance for physics.
These included the discovery of the neutron, which put nuclear
physics on a sound footing for the first time; the discovery of the

positron, the first evidence of antimatter; the publication of Werner Heisenberg's seminal papers consolidating the theory of the nucleus; and key breakthroughs in the development of particle accelerators. When Serber received his Ph.D., in 1934, some of the most important theoretical work in exploring the vast new world opened up by these events was taking place at the University of California at Berkeley around a charismatic young theoretical physicist named J. Robert Oppenheimer. Later in 1934, Serber had the good fortune to run into Oppenheimer at a famous summer school in Ann Arbor, Michigan, the good judgment to rearrange his plans to go work with him, and the acuity to become Oppenheimer's principal research assistant. Joining Oppenheimer, Serber now fell in with yet another intense, intellectually active group.

At the time, physics was not as specialized as it would be only a few years later, but was rather one vast frontier without fixed boundaries in which pioneers could roam and explore at will. As a postdoc at Berkeley under Oppenheimer (1934–1938) and then as a professor at Urbana (1938–1942), Serber's work overlapped domains now called high-energy physics, cosmic ray physics, nuclear physics, cosmology, and accelerator physics.

In December of 1941, within hours after Oppenheimer was informed that he would soon be named head of a laboratory to build the atomic bomb, Serber was Oppenheimer's first recruit. Serber's wife Charlotte headed the Los Alamos library, and was the only female group leader in the Manhattan Project. In late July 1945, a week after the Trinity test in Alamagordo, New Mexico, the first explosion of an atomic bomb, Serber went to Tinian Island, the take-off point for the bomb delivery planes. After the Japanese surrender, Serber went to Japan and inspected Hiroshima and Nagasaki as head of the first team to survey the (nonmedical) effects of the bomb.

After the war, Serber was a professor of physics at Berkeley (1945–1951), and then a professor at Columbia (1951–1978). In July 1948, following anonymous accusations, he was subject to an Atomic Energy Commission investigation into his "character, associations and loyalty." This was at the beginning of Cold War paranoia, though two years before the rise of Sen. Joseph McCarthy and six years before the Oppenheimer loyalty trial. It was no surprise that the FBI had files on the more politically active members of the Leof family, such as Madelin Blitzstein, her husband Sam, and her stepchild Marc Blitzstein. But Serber himself had never been in any sense a political radical or activist. Both as a member of the Leof clan and of the Oppenheimer circle, Serber was at most an often sympathetic observer of a sometimes politically naive crowd. Yet that much was sufficient to condemn him in some circles. His loyalty, and that of his wife Charlotte, had been questioned (unbeknownst to them) even as they were working at Los Alamos. The postwar charges also may have been at least partly motivated by Serber's efforts to keep politics out of discussions of the technical feasibility of early designs for a hydrogen bomb. Serber ran afoul of Edward Teller, most notably, in his efforts to separate technical and political considerations, a separation that Serber viewed as essential to protect the integrity of science.

Serber also served as a consultant and adviser for numerous laboratories, including Brookhaven National Laboratory, Argonne National Laboratory, the Fermi National Accelerator Laboratory, the Stanford Linear Accelerator Center (SLAC), and the Los Alamos Meson Physics Facility. Following Oppenheimer's death in February 1967, and of Charlotte's that May, Serber became a companion of Kitty Oppenheimer and helped her arrange an annual conference in memory of her late husband. In 1971–72 Serber was president of the American Physical Society (APS). Anti-Vietnam War and anti-administration protests by APS members happened to crest during his term as president, and activist members tried to have the organization, whose charter restricted it to scientific concerns, take stands on political issues. Viewing this as another kind of threat to the integrity of science, Serber opposed these efforts, and now found himself in the unfamiliar role of principal figurehead of the old guard.

In 1974 Serber became a trustee of Associated Universities, Inc., the institution that runs Brookhaven, at a time when it had just embarked on building a major new accelerator called ISABELLE, and in 1975 he became chairman of Columbia's physics department. He stepped down as physics department chairman at the mandatory retirement age of sixty-eight in 1978 and as AUI trustee in 1981, only one year shy of the fiftieth anniversary of the publication of his first paper. The end of Serber's professional career coincided with the end of the "golden age" of U.S. physics. His retirement as department chair occurred at just the time the standard model of elementary particle physics became firmly established, able to explain virtually all data obtainable on existing instruments. And his retirement as AUI trustee was just two years prior to cancellation of the ISABELLE project, an event that in turn anticipated the cancellation of the Superconducting Super Collider a decade later, and a warning sign of the end of the near-total dominance that the United States had held over accelerator physics since the 1930s.

The following is Serber's memoir of the above events, a project that arose out of his Pegram lectures in 1994 at Brookhaven National Laboratory. The Pegram lecture series, established in 1959, is given annually "to provide an opportunity for distinguished scholars to examine the interaction between science and other aspects of our culture and society." Serber devoted his to a recollection of the Los Alamos years, of his missions to Tinian and Japan, and of the immediate postwar years at Berkeley, Columbia, and Brookhaven.

According to a long-standing agreement, the Pegram lectures are ordinarily submitted to Columbia University Press for possible publication. Serber's lecture notes were insufficiently developed for a book, and as Brookhaven's historian I was asked to help Serber in producing a full-fledged memoir. My assistance took the form of incorporating into his Pegram notes other historial material Serber had published, along with new material drawn from interviews and letters, and encouraging him to extend his reflections to issues and people he had not mentioned. From these, I prepared a manuscript for him to correct and revise. The style and content of the memoir thus remain Serber's own. A few days after going over the final draft of the manuscript, he underwent surgery to remove a brain tumor. He never fully recovered from the operation, and died on June 1, 1997, at the age of eighty-eight.

To fully appreciate the nature and originality of Serber's scientific contributions, his work would need to be discussed alongside a history of physics—for instance, the excellent and extensive three-volume *Twentieth-Century Physics*.[2] For Serber's specific contributions to the theory of nuclear forces, one should consult *The Origin of the Concept of Nuclear Forces*; for a history of Los Alamos during the wartime years, see *Critical Assembly: A Technical History of Los Alamos During the Oppenheimer Years, 1943–1945*.[3] But since these memoirs contain much of value to readers interested in history who lack training in physics, it may be worthwhile to sketch out, however briefly, some of the scientific background.

When Serber first began working with Oppenheimer in 1934, the latter was wrestling with the problem of the self-energy of the electron, or the way that the electron interacts with itself, and the vacuum, due to its electric field. As Oppenheimer (among others) had noticed in the first paper (1930) he had written after returning from Europe to the United States, when quantum electrodynamics, or the theory describing the interactions of electrically charged particles and fields, was applied to the self-energy problem, the theory gave nonsensical infinite results. Throughout the decade, Oppenheimer was convinced that quantum electrodynamics was fundamentally flawed and eventually would be replaced; still, he and associates strove to find ways of using the theory to describe self-energy. Self-energy would turn out to involve three principal pieces. One involved the straightforward interaction between an electron and its own field; a second involved a "vertex correction," or what happened in cases following deflection of the particle through, say, interaction with a photon; and a third involved polarization of the vacuum, or the effect of the electron's charge on the pairs of virtual particles that constantly surround a particle. Under Oppenheimer, Serber tackled two of these three major pieces (papers 7, 8 in the list of Serber's writings at the end of this volume). In one of these papers, Serber used the word "renormalize" for apparently the first time in print to describe the theoretical strategy by which the divergences are removed, a term that has been retained until this day. The successful renormalization of quantum electrodynamics, however, took place only after World War II.[4]

In the 1930s the theory of quantum electrodynamics was intertwined with the understanding of cosmic rays.[5] By then, much experimental information about cosmic rays had been collected, and the prevailing assumption was that primary cosmic rays consisted of electrons and positrons, which when passing through the atmosphere produced high-energy photons, producing more electron-positron pairs, and so forth, as well as "knock-on" electrons that had been struck by other particles, resulting in an eventual shower of particles at the earth's surface. But how to apply the theory of quantum electrodynamics to this information was unclear. One puzzle was the existence of a "penetrating component" in such showers; a portion of the cosmic rays was more able to pass through matter than ordinarily would be expected from electrons and positrons. Was this an indication of the breakdown of quantum electrodynamics at the high energies involved, or of something else? In 1937 Oppenheimer and Serber were the first in the West to propose that a new particle found in the penetrating component, called a pi meson or pion, might be the same as a particle proposed by Hideki Yukawa to explain the existence of the nuclear force (paper 12). They were also the first to draw the attention of physicists in the West to this seminal work by Yukawa, who in 1949 would win the Nobel Prize for it. But Oppenheimer and Serber were also the first to have misgivings about this identification, pointing out some difficulties that would result, and the identification indeed was discovered to be erroneous a decade later.

Yukawa's suggestion occurred at yet another intersection of fields: cosmic rays and nuclear physics. He had proposed his particle in attempting to explain where nuclear forces come from—the most basic question of nuclear physics—and his proposal gave rise to many theories about pions and how they might produce nuclear forces. Serber devoted much of his work to pion physics and he made the earliest serious attempt in the United States to describe nuclear forces along the lines that Yukawa had outlined (paper 15), discovering more problems with the identification of the new cosmic ray particle with the one involved in nuclear forces. Serber's work on pion physics continued after the war, when the new accelerators advanced nuclear physics by leaps and bounds, making it possible to carry out experiments in high-energy neutron and proton scattering. Among his other contributions during his Berkeley years, Serber helped to pin down the zero spin of pions, and he also helped to establish the fact that pions were actually being produced by Berkeley's 184-inch cyclotron, for the first time by an accelerator.

Another Serber contribution to the theory of nuclear forces was to understanding its "shape," or dependence on the distance of separation. Initially, the nuclear force was conceived to be simple and monotonic, modeled on the electromagnetic (Coulomb) force. This approach, however, encountered difficulty explaining the "saturation" of particles in the nucleus, or why they are not packed tighter and tighter the closer they are to the core; rather, they simply occupy more volume when added to each other, the way molecules of water do. Heisenberg had considered, but rejected, the idea of a "repulsive core," or a force that was repulsive at small distances and attractive at larger distances. While at Berkeley, Serber rehabilitated this idea and showed how it could be used to explain the new scattering data being obtained from accelerators; for a while it was even called the "Serber force." Finally, Serber also pioneered the theory of high-energy nuclear physics, a completely new field opened up by particle accelerators, and Serber's work embodied a new set of ideas about nuclear behavior that led to the creation of what became known as the "cloudy crystal ball" model of the nucleus.

Serber also contributed to a major step forward in nuclear reaction theory (papers 13, 14, 17). This work derived from a proposal by Heisenberg in his 1932 papers on the theory of the nucleus. Heisenberg had treated the proton and neutron as symmetric, that is, as different states of the same particle. Formally, Heisenberg represented this situation analogously to the way spin was represented— by treating the particles as spinning on an imaginary axis, with particles being protons or neutrons depending upon whether the axis was pointed up or down. This was enormously convenient because the particles could then be treated as substates of a quantity which came to be known as isotopic spin. In 1938 Oppenheimer and Serber took a huge step forward in the theory of nuclear reactions. Following up on some published work and a suggestion by Gregory Breit, they proposed that isotopic spin was conserved in nuclear reactions; that the total amount remained the same, although the symmetry sometimes may be only approximate due to the influence of factors such as the Coulomb force.[6] Oppenheimer and Serber then used this idea to develop a theory of nuclear transitions that allowed them to predict which ones would be permitted and which ones forbidden. The work was enormously significant in paving the way to the future, for establishing symmetries that could lead to "selection rules" was a philosophy that was to become a fundamental procedure of nuclear and particle physics. The Oppenheimer-Serber paper also drew the attention of theorists to the importance of studying analogue states and mirror nuclei, that is, different isotopes with analogous isotopic spin.

Astrophysics and cosmology was another interest of the members of Oppenheimer's circle, and Serber participated in an unsuccessful attempt to work out the energy-producing nuclear cycle inside stars, which was successfully carried out later by Hans Bethe. Serber and Oppenheimer also worked on stellar cores (paper 19), and Serber was present at the initial discussions between Oppenheimer and Snyder which ended up in a paper that first predicted the existence of black holes.

In accelerator physics, Serber worked with Don Kerst on the early development of the betatron, in what is probably the first time the theory of electron orbits was worked out in detail before the accelerator was actually designed. Serber also worked with Ernest Lawrence's crew on the theory of the 184-inch synchrocyclotron.

An independent thinker, and unaffected by the fashions of the moment, Serber was one of the few theorists who continued to work in field theory throughout the 1960s, a time when it was fashionable to do other things like nuclear democracy (the idea that each particle was somehow composed of all others). This contributed to Serber's role in assisting Caltech theorist Murray Gell-Mann in the invention of quarks.

But Serber's full impact on physics is not revealed by his published writings alone. One reason is that he was an innovator uninterested in staking claims for priority. Serber was content once he had solved a problem and passed along the solution to someone who could use it. In many papers issuing from Brookhaven in the 1950s, for instance, footnotes thank Serber for key contributions which Serber easily could have published separately had he been so inclined; the Pais-Piccioni paper on the "regeneration" effect is one example; so is the Landé, Lederman, and Chinowsky paper on the discovery of the K-long particle.[7] But an even more important source of influence was Serber's ability to survey the state of theoretical physics in a way that others found easy to understand; he understood, and was able to articulate, how it all fit together. Again and again throughout his career he served as a coordinator for experimenters and theorists alike in elaborating a vision of how high energy or nuclear theory worked as a whole, helping colleagues to spot inconsistencies in the theory and pushing them to investigate areas where it was wanting. According to Joseph Weneser, a theorist who arrived at Brookhaven a year after Serber:

> Serber liked to tackle complex issues and think them through until he had discovered the simple principles behind them. Then, when he'd explain them to you, he'd make it seem as easy as "What goes up must

come down," and you'd wonder why there ever was a problem. And he was always available for discussions. Those kinds of people— Fermi was another—have a tremendous influence on the field, which may not come across just from papers or citations.

Serber performed this synoptic service in many different contexts. At Berkeley from 1936 to 1938, as Oppenheimer's research assistant, he frequently had to interpret Oppenheimer's often obscure remarks to students he was advising on a wide range of areas in physics. In July 1942, on leave from Urbana, Oppenheimer gave Serber the task of leading off the discussion about the overall prospects for the feasibility of an atomic bomb, at a meeting in Oppenheimer's Berkeley office. In March of 1943, right after the opening of the still-incomplete laboratory at Los Alamos, New Mexico, Serber delivered a series of five lectures summarizing the state of knowledge about the still-theoretical bomb, lectures that were declassified in 1965 and published in 1992, with Richard Rhodes, as *The Los Alamos Primer*.[8] These lectures are, no doubt, Serber's most important single contribution to the bomb effort. After the war, at the Berkeley Radiation Laboratory, Serber was asked to review the state of the field for the benefit of the experimenters, and he would perform similar services at Brookhaven, Fermilab, and SLAC.

In Nuel Pharr Davis's book *Lawrence and Oppenheimer*, the author provides the following description of Serber at his Los Alamos lectures:

> A lean, dark, inconspicuous wisp of a man, Serber hated dramatics. While talking he stumbled continually over his words and swallowed impatiently as though he had dust from the mesa lodged in his throat but felt his subject too trivial to justify a sip of water. Nevertheless, he held his audience. "He wasn't much of a speaker," says one of those present. "But for ammunition he had everything Oppenheimer's theoretical group had uncovered during the last year. He knew it all cold and that was all he cared about."[9]

As this description suggests, Serber's style was to observe an entire complex system and how its various parts fit together, and then to express this informally, economically, and without fanfare. Yet if Serber was an innovator uninterested in priority, he was also a teacher uninterested in textbooks. When his Berkeley lectures were written up by a colleague, they were called simply *Serber Says: Volume I* and *Serber Says: Volume II*. "Serber Says," indeed, became a kind of trademark title for Serber's synoptic talks, expressing at once their comprehensiveness and informality. His APS presidential address was entitled "Serber Says, Vol. III," and, in 1987, a collection of his lectures on nuclear physics was published under the title *Serber Says: About Nuclear Physics*.[10] His approach was well suited to colleagues and graduate students; in his entire postgraduate career, he never taught an undergraduate course.

There is no such thing as *the* scientific personality, nor even the physics personality. Anyone with the slightest acquaintance of the field of physics is aware of its colorful range of practitioners who work in a variety of ways to bring different temperaments and skills to bear on a choice of problems to tackle and the ways they tackle them. Serber was a distinctive figure among his colleagues in his ability to grasp a variety of details in their complexity, integrate them through the whole, and then explain it all in simpler terms than he found it. In sharp contrast to many of his colleagues, Serber was shy, unostentatious, and willing to stick with what he believed and defy even the most pervasive fads. In the sumptuous banquets of the Columbia University Physics Department's renowned Chinese lunch—at which most people let faculty member T. D. Lee order exotic dishes for them—Serber was the one who could be perfectly happy, and unembarrassed, with a plate of chow mein. He *liked* the role of the relatively detached spectator. One of the few subjects about which I saw him get visibly agitated was the Copenhagen Interpretation of quantum mechanics; that the observer plays a role in the reality of the world. Serber vociferously objected. Physics was about laws of nature that had been around since the beginning of the universe, he would insist, laws that were independent of human beings. But while he talked several times about inserting a discussion of this subject into the manuscript, he never found a way to do so that satisfied him.

Many events in the following memoir are related in spare and sometimes almost terse prose, employing well-served understatement, a keen eye for the significant detail, and a dry wit. Serber relates coolly and dispassionately events that by others might have been rendered more judgmentally or with greater emotional affect: the *coup de foudre* between Oppenheimer and Kitty Puening; the security hearing at which Serber's loyalty was questioned; a colleague propositioning his wife; his wife's suicide; Kitty's painful death during an around-the-world sailing excursion. In describing an accident Charlotte had while horseback riding, which left her on the ground in a blood-soaked jacket with a fountain of blood pouring out of a pierced artery, Serber carefully pauses to note the reactions of the others present—Robert Oppenheimer, Frank Oppenheimer, and Ed McMillan—and how these reactions expressed the personalities of each. Serber's descriptions can frustrate. Some readers may wish that Serber, one of Oppenheimer's closest associates, had provided more insight into this fascinating personality, something that might help elucidate Abraham Pais's remark that "vast insecurities lay forever barely hidden behind his charismatic exterior, whence an arrogance and occasional cruelty befitting neither his age nor his stature." One is also left wishing for a more nuanced view of Kitty Oppenheimer, a fascinating and difficult person in her own right (Pais called her "the most despicable female I have ever known, because of her cruelty"), with whom Serber lived after the deaths of their respective spouses.[11] Throughout all the wide variety of situations which he describes and people with whom he interacts, his tone remains highly consistent within a narrow bandwidth.

The most riveting passages are those dealing with Serber's wartime experiences in Los Alamos, Tinian, and Japan. Serber wrote Charlotte regularly after he left Los Alamos for Tinian (it was the first time the two had been separated for any length of time), and large passages of these letters are quoted here. Serber describes what he saw (though for the first several weeks the letters were written under the constraint of military censorship). As always, he passes judgment neither on the people he meets nor on the situation he is in, nor on himself and his own actions. Though not without emotion, these descriptions lack the familiar tone of foreboding and moral condemnation of much writing about what may well be the most highly charged and ethically controversial event of the century. In some ways this makes his descriptions even more effective, for the familiarity and reassurance of moral condemnation can anesthetize the tragedy by setting us at a (superior) moral distance. Here, too, Serber expresses himself simply and dispassionately, sprinkling his prose with telling details and dry humor. Thus he writes of the brilliant colors of the coral in the waters off Tinian; of the particular way the B-29 bombers are stacked on the runway "like cars coming back to a city on a Sunday night"; of the heavy iron office safes that alone protruded above the rubble of Hiroshima; of the grazing horse whose hair had been burned on one side by the fireball and not on the other. It is interesting that Serber notes Gen. Curtis LeMay's decision to allow planes to be so heavily loaded with bombs that some crashed on takeoff because it saved American lives in the long run, because this reasoning seems to have mirrored Serber's own with respect to the decision to use the atomic bomb.

An excerpt of this section was published in *The Sciences*, published by the New York Academy of Sciences, in the summer 1995 issue, at the time of the fiftieth anniversary of the atomic explosions over Hiroshima and Nagasaki. Anticipating that many would find the tone of these passages perplexing and even disturbing, I drafted for Serber an outline of a possible moral justification of work on the bomb project, and asked him to set down his thoughts for inclusion in the article. "It is an important point," I wrote, coaxing him to publicly commit himself. "It *is* an important point," he replied, "and I think it's served well by understatement." As a result, the excerpt appeared essentially as it is here.

Some individuals indeed had negative reactions to the article. One wrote that the article was composed "so prosaically and without emotion that my immediate reaction was to recall Hannah Arendt's famous phrase, 'the banality of evil.'" Another condemned Serber for failing to confront "the enormous human tragedy." Still another announced that he was resigning from the New York Academy of Sciences in protest. To the Japanese people Serber resembled an "SS man working for Adolf Hitler," the correspondent wrote, declaring that the "truth" about the dropping of the bomb over Hiroshima was that Truman was attempting to intimidate Stalin.[12]

As these letters show, the U.S. effort to build an atomic bomb during the Second World War has become a kind of ethical Rorschach test, in which judgments expressed tend to reveal the assumptions of the people who render them. These correspondents assumed that only the suffering of the Japanese victims of the bomb should be factored in to moral thinking—and not that of, say, victims of Japanese actions that might have been forestalled by the bomb. But the letters also reveal an even more fundamental assumption about the nature of moral conduct itself—namely, that exercising restraint regarding expression of one's moral thinking means the absence of moral responsibility.

To modern sensibilities, a much more acceptable reaction was Oppenheimer's public breast-beating, his recollection of Vishnu's apocalyptic saying, "Behold, I am become death, the destroyer of worlds"; and his remark that "physicists have known sin." In contrast, Serber's restraint is unsettling. Our age, so distant from the reality of all-out warfare, so convinced that it can find simple truths in complex and ambiguous historical events, and so confident that it occupies high moral ground, finds it hard to accept an action as ethical unless it is accompanied by full disclosure of feelings and motives. Keeping one's feelings and moral thinking to oneself is viewed as taking a moral version of the Fifth Amendment: an implicit admission of guilt, or a lack of moral thinking altogether, the result of conformity and a retreat into denial.[13]

Science historian Spencer Weart has observed that, according to (false) popular mythology, all of those involved with the Manhattan Project feel guilt and remorse. Ever since the mid-1950s, he notes, Paul Tibbets, the pilot of the *Enola Gay*, the B-29 that dropped the first atomic bomb over Hiroshima, has received telephone calls from reporters on the anniversary of the bombing, curious about his sanity. "The reporters seemed disappointed when they learned that Tibbets and his crew were all leading ordinary lives with no regrets."[14] What was true of Tibbets is even more true of the Manhattan Project scientists; it is as if their participation in the event ipso facto turned them into public figures, of whom we expect at least voluntary acts, or involuntary behavior, by which they symbolically shoulder remorse on behalf of all humanity.

Serber put down all he felt was sufficient on this issue. (Another contributing factor may have been that it was out of character for him to express himself on a subject where he might wind up repeating what others have said better.) More protracted discussion may seem unnecessary: to some, it will appear a superfluous justification of reasonable conduct; to others, an attempt to defend the indefensible. But I believe the issue raises deep moral questions that are worth pursuing. Those who would fault Serber for failing to live up to modern moral expectations ought at least to examine these expectations themselves. Do participants in an event with extraordinary human consequences have a social responsibility to air their consciences in public? Is it morally illegitimate for them to be discreet with respect to their reasons and feelings? Do people who write about such an experience have an obligation to be more than dispassionate? Modern sensibilities intuitively assume the affirmative. But the fact that that answer is at least debatable shows that Robert Serber's restraint, far from revealing the absence of intelligence in a moral framework, challenges us to reexamine our own expectations about it.

Many scientific memoirs are used to claim credit or settle scores, but Serber's is not one of them. When Serber talks of having a hand in discoveries where credit is generally given to others—for example in Murray Gell-Mann's invention of quarks—Serber is merely clarifying a role that Gell-Mann himself has already acknowledged. And if Serber has a score to settle, it would be with Edward Teller. Teller not only altered a document relating to the feasibility of H-bomb development on which Serber had collaborated—to make it much more optimistic—but was the "confidential informant" who made the preposterous statement to the FBI that Serber, along with Oppenheimer and Philip Morrison, was considered "one of the most extreme leftists among physicists."[15] (Serber's activism was so mild that he was willing to sign the controversial loyalty oath imposed by the Regents of the University of California because he did not take it that seriously.) Yet Serber, with what can only be described as moral character, continued to treat Teller as a good friend to the end of his life.

Many scientific memoirs, too, begin with a dramatic scene in the life of the author; Serber's colleague Luis Alvarez, for instance, begins his memoir with the story of his flying the Hiroshima mission, before backtracking to his youth. Serber might have used a number of episodes for this purpose, including his own tale of getting kicked off the Nagasaki mission. But he adamantly refused. Consistent with his concern for order and integrity, he insisted on beginning with the question of the date of his birth. From the opening paragraph, he reveals how important it is for him to speak only what he knows to be the case and nothing more, and his bemusement at those who are satisfied with less rigorous and merely formalistic criteria for truth. If what one hopes from a memoir is a sense of the person, then this one serves admirably well.

Acknowledgments

Joe Anderson, Henry Barnett, Shirley Barnett, Sam Bono, Patrick Catt, Walter Chudson, Fred Knubel, Roger A. Meade, Priscilla McMillan, Joe Rubino, Robert Sanders, Linda Sandoval, Irene Tramm, Georgia Widden, and Lynn Yarris assisted us in tracking down information and pictures. Laurie M. Brown, Gerald Lucas, John H. Marburger, Stephanie L. Stein, and Joseph Weneser provided helpful comments on the introduction and various parts of the manuscript. We owe a great debt to Roy E. Thomas for his work in editing the manuscript and to Ed Lugenbeel for steering the book through the press. We are grateful to Brookhaven National Laboratory for supporting the project, and to Michael Tannenbaum for bringing the two of us together on it. Excerpts from "Peaceful Pastimes: 1930–1950," by Robert Serber, are reproduced with permission from the *Annual Review of Nuclear and Particle Science*, vol. 44, copyright © 1994 by Annual Reviews, Inc. Excerpts from "Eyewitness to the Bomb," by Robert Serber with Robert P. Crease, are reproduced courtesy of *The Sciences* (July–August 1995). Use of interviews from the Oral History of Robert Serber by Charles Weiner and Gloria Lubkin on February 10, 1967, courtesy of the Niels Bohr Library, American Institute of Physics, College Park, Maryland.

Peace and War

PART I

Peace

ONE

Philadelphia and Madison, 1909–1934

I was born in Philadelphia on March 14, 1909—an alleged fact that the FBI was unable to verify years later, in 1942, when they were processing my clearance for top-secret work as Robert Oppenheimer's principal assistant in the initial study of the feasibility of an atomic bomb. They couldn't find the birth certificate. A security officer at the Metallurgical Lab of the Manhattan District in Chicago, which would formally be my employer, when faced with this difficulty put my papers in his desk drawer and forgot all about them. I had been working on the project at Berkeley for six months before anybody noticed that I didn't have a clearance. When the security people told me that the problem was a missing birth certificate, I suggested that they get in touch with my sister Alice, who, though younger than I, knew much more about the family history. Alice went out and found the doctor who had delivered both her and my younger brother. Counting on age, senility, and the poor record-keeping prevalent early in the century, she told him he had also delivered me. He signed an affidavit to that effect, on the basis of which a delayed birth certificate was issued and my clearance was completed.

When I became sixty-five and eligible for Social Security, I presented this certificate to the Social Security Administration. They apparently had tougher standards than either the FBI or Army Intelligence, because they did some additional checking, and a year later they sent me a photostat of records of the Census Bureau's 1910 census in Philadelphia. It listed "David and Rose Serber and infant son." The March 14 has to be taken on trust.

Fig. 1.1 David and Rose
Serber's wedding picture
around the turn of the century.

At the time my parents lived in West Philadelphia. The first house I remember was on 41st Street near Girard Avenue and close to Fairmount Park. As a child I spent a lot of time wandering around the park—along the banks of the Schuylkill River and in Horticultural Hall and Memorial Hall, both of which had been built at the time of the Centennial celebration in Philadelphia in 1876. My grandfather Serber, a man with a long, flowing white beard, lived a few doors away. When I was five or six, he used to give me a nickel every Saturday so I could go to the movies around the corner, where the standard fare was serials such as *The Perils of Pauline* starring Pearl White. My grandfather had immigrated from Russia about 1886 when my father was two years old. I heard a story that he gained the name Serber at Ellis Island as an abbreviation of something longer. My mother's family came from Poland about the same time, but she was born in Philadelphia.

Later, around the time I first started going to school, we moved around the corner to a house on Girard between 41st and 42nd. In those days, horses were still a common part of everyday life, and to hitch a ride to school we would jump on the back step of a horse-drawn ice wagon. I'd stop off on the way at a little store, where for two cents one could buy a soft pretzel to be eaten at recess time. I went to a public school, the Leidy School, where a classroom held one teacher and supposedly not more than fifty students, though there were usually one or two extra. The only out-of-the-ordinary curriculum that I remember came each Memorial Day, when the school was visited by veterans of the Civil War, all dressed in their blue uniforms.

I remember little of my mother, Rose Frankel. My clearest recollection of her is one night when I was about four, when she came into my bedroom to kiss me goodnight before leaving for the theater. She was all dressed and made up, and I thought she was beautiful. When I was about six, she came down with a disease of the nervous system. Later, we moved to Atlantic City for a year because the sea air was supposed to be easier on her, and I spent another year in Philadelphia living with an aunt while my parents remained in Atlantic City. During that time I came down with a bad case of scar-

Fig. 1.2 Rose Serber and Robert.

let fever which lasted two months. It left me nearsighted and I had to wear glasses after that. After I moved back with my parents I practically never saw my mother. She couldn't stand noise or light, and thenceforth I only saw her once or twice a week in a darkened room. We children were habitually being shushed and warned to be quiet so as not to disturb her.

Beginning at age seven, I would spend summers at Camp Arcadia, a boys camp in Maine. My favorite activities were baseball, swimming, and sailing. I wasn't very good at baseball, but the last year I won the gold medal in the camp swimming competition. The camp had a number of sail canoes and a small sloop, and it was there that I learned to sail. I was there with my brother William in the summer of 1922 when we received a surprise visit from my father. He had driven from Philadelphia to tell us of our mother's death. I remember our crying at the news.

My father was a lawyer. A pleasant, agreeable man, he was interested in literature and politics and active in the reform movement of the local Democratic party. When I was just barely old enough to read, I remember seeing posters on all the telephone poles in the neighborhood with his picture on them, saying that he was running for City Council. He told me he wasn't expected to win, but that his candidacy would bolster the chances of the rest of his ticket in our district. I remember once—it must have been in 1915 when I was six—when he tried to explain to me who the good and bad guys were in the World War, and how embarrassed I was at the end when he quizzed me and I got it wrong. A year later, in 1916, I remember one November morning his surprise when, on the telephone, he got the news that Wilson was elected, whereas the night before he told me it was Hughes; he explained that the California vote coming in late at night had upset the election. When I was in high school, we lived in a hotel he owned, the Bartrum, at Woodlawn and Walnut, right next to the University of Pennsylvania campus. We learned good table manners; we ate in a formal dining room, served by waiters in tails. At that time my father was the American attorney for Amtorg, the Russian trading company, and we were served fresh caviar every morning for breakfast.

My sister Alice is two years younger than I, and my brother William one year younger still. Alice quit the University of Pennsylvania when in biology lab she was required to dissect a frog—then, surprisingly, went on to become a medical technician. She married an artist, Robert Carlen, who was a conventional starving artist until he decided, simultaneously, that he wanted to marry Alice and that he would never become a great painter. He had excellent taste and a thorough knowledge of art. He went to an auction at the Philadelphia Customs House where they disposed of unclaimed and abandoned articles, bought a trunk of German prints for $200, and turned around and sold them for $10,000. With this estate, he very successfully established a family and an art gallery. He is notable as the discoverer of most of the known paintings by Edward Hicks. His Hicks painting of William Penn making a treaty with the Indians hung on loan in the Treaty Room of the State Department for many years.

William became a radiologist. He married Jane Greenberg, the daughter of a prominent Philadelphia realtor. During the Second World War he served in the Army medical corps, and afterwards specialized in cancer therapy at a number of Philadelphia hospitals. For years he was chief of radiation therapy at Philadelphia General Hospital. When that closed down, he went to Hahnemann, where he was on the staff of the radiation oncology department. He tried retiring in 1996, but gave up after three weeks. He is still practicing at the age of eighty-four.

When, in 1922, I entered Central High School, on Broad Street near Spring Garden, I had the idea of becoming an engineer. My uncle, Lester Goldsmith, was chief engineer for the Atlantic Refining Company, and he steered me in that direction. I didn't take the regular academic sequence but rather what was called "Industrial Arts." This was a more vocational sequence of study, with courses in drafting, blacksmithing, pattern-making, cabinet-making, and machine shop, but also with a good deal of physics, chemistry, mathematics, English, and French.

At the time, Central High was an exceptional Philadelphia institution. Its students came from all over the city, and its standards were more that of a community college than a high school. On graduation, students in the academic course received a BA degree, and those in industrial arts a BS. The science teachers were mostly connected with the Franklin Institute, and were actually competent. The physics course was easily the equal of standard college courses.

I had two particular friends in high school named Al Paris and

Harry Zimmer. Al was French; his family had come to this country only recently, and after he graduated from Central High he and his family returned to France. I had a letter from him after the war, saying he had survived it and was living in Lille, was married, and had a college-age son. Harry Zimmer entered Lehigh the same time I did, but we drifted apart. After the war I heard that he had been a colonel in the signal corps and was killed on one of the islands in the Pacific.

During my freshman year, I flunked gym because of too many absences. I had to come in during the afternoons for makeup time. During the makeup, one of the instructors there got me interested in gymnastics, and I began to learn all the standard moves. Harry Zimmer was also interested, and together we joined a German gymnastics club, the Philadelphia Turngemein, and spent two or three evenings a week in gymnastics classes. I got to the point where I could do a reasonably good giant swing. Years later, at a New Year's Eve party in Berkeley after the war at Frank Oppenheimer's house, Luis Alvarez and I discovered we had both been gymnasts. This was after a few drinks, and we decided that a structural iron bar across Frank's living room would make a fine horizontal bar. Much to the amusement and alarm of the other guests, we tried to exhibit our skills.

I was also on the school's swimming team. Diving was my favorite event, but I didn't quite qualify for that and ended up doing

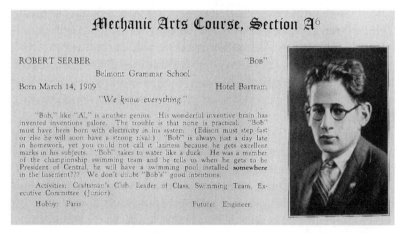

Fig. 1.3 From the Central High School yearbook, 1926; "Al" was Al Paris, a boyhood friend.

the breaststroke at 100 and 200 yards. In my senior year, the Central High swimming team won the National High School championship, though my contribution to that achievement was minimal and I didn't even go to the national meet.

As soon as I was sixteen I got a driver's license, and for the next few years was a very reckless driver. I tried never to let another car beat me away from a traffic light. In retrospect, I'm surprised I never got into a serious accident. My father bought a Chrysler the year after they first came out; they were considered a very sporty car then. Our license number was A14. In those days, a special number couldn't be bought but was a sign of political pull, which had the disadvantage that it was easily remembered by offended onlookers. Once, when I was sixteen and my cousin David, who was a year younger, was visiting, he asked me to teach him to drive. We found an empty-looking road just over the city line, and he took the wheel. A couple of minutes later, ahead of us, we saw a motorcycle cop standing at the side of the road next to his parked machine. David said, "What should I do?" I said, "Keep going." Of course, he had no learner's permit and was terrified. He kept his eyes on the cop, with the result that he sideswiped the motorcycle. Looking in the rearview mirror, I saw the cop with his mouth wide open, and his motorcycle lying in the ditch. David said, "What should I do?" I said, "Step on it!" I reached my foot around his and pressed the accelerator down all the way. For some reason—perhaps his machine was out of commission or he simply had compassion for two kids—the cop didn't pursue us.

I graduated from Central High in 1926. Following my uncle's advice, I went to Lehigh University, in Bethlehem, Pennsylvania, a good engineering school. As a student, I was quiet and kept to myself, and was often unsophisticated and naive. On the first day of classes, I read off my chemistry lab assignment of room and desk from a notice on a bulletin board, went there, and found the room completely empty. That I thought a little strange, but the apparatus was all there, I had my lab book, which told me what I was supposed to do, so I obediently got to work. The same thing happened the second day, and continued for the rest of the semester—nobody else ever showed up, I always worked by myself, and I never questioned it. On the last day of classes, after finishing my assignment, I decided to look around. In the next room, I found a laboratory bustling with activity. I went over to the instructor and handed him my stack of the

Fig. 1.4 The young scholar at his desk, during his college years (late 1920s).

semester's lab reports. He was astonished to see me and demanded to know where I'd been all semester. I was supposed to be with everyone else, of course, and evidently the initial bulletin board posting had been incorrect, but I was the only one who didn't have the wit to realize it. He shook his head in disbelief when I told him, but accepted my lab reports. They must have been all right, because at the end of the freshman year, I won the prize for having the highest grade average in the freshman class. The prize was a slide rule in a handsome brown leather case, on which they had printed my name in large black letters. I still have it on my desk: I used it to do calculations for a good many years and took it to Los Alamos, though by that time we were using Monroe desk calculators for our important problems.

At the end of the freshman year at Lehigh, engineering students had to specify which branch of engineering they wanted to major in. I had intended to follow my uncle as a mechanical engineer, but I found I liked the pure science courses better than the engineering courses. A new curriculum, called Engineering Physics, had just been established the year before, and I selected that. For one thing, it had the advantage that, being new, the requirements were not too well established, and there were not too many required courses. I was free to concentrate on what interested me most—which was physics, chemistry, and mathematics. We went through the standard physics curriculum—mechanics, light, heat, sound, electricity, and magnetism—and ended up in my senior year in a course in atomic physics. I suppose it included the old quantum theory—Planck, the Bohr atom, and the photoelectric effect. The new quantum mechanics—the work of Dirac, Heisenberg, and Schrödinger—was only three years old when I was a senior.

During my freshman year, I was on the swimming team, but I didn't continue after that. Bethlehem was only sixty miles from Philadelphia, and thanks to the Reading Railroad it was easy to get back home. As a result, I didn't have much of a social life in Bethlehem itself. About the only interesting events I remember arose from the fact that the father of one of my classmates ran a gambling house in Bethlehem for the executives of Bethlehem Steel Company, and we used to go there for lunch, where we enjoyed caviar and Bermuda onion sandwiches on rye.

While I was in high school, my father had remarried to Frances Leof, a ceramicist, muralist, and potter. An uncle of my stepmother was a prominent Philadelphia doctor, a general practitioner named Morris V. Leof. Leof was a charismatic man who had been born in Russia and immigrated to the United States before the turn of the century and who had a four-story brownstone house at 322 South 16th Street. His professional offices were on the ground floor, a large living room and dining room and kitchen were on the second floor, and the bedrooms were on the two upper floors.

"322," as we called it, was a kind of salon for young writers and artists of Philadelphia, mostly because of the interests of the Leof children, but perhaps partly because Mrs. Leof—Jenny—always had the dinner table set for twelve, just in case someone dropped in for dinner. The oldest of the Leof children was Madelin ("Madi"), a freelance journalist and aspiring novelist. She had the remarkable ability to type an article and carry on a conversation about an entirely different subject at the same time, an art later known in computer language as time-sharing. She was born in 1901, and in 1928 married Sam Blitzstein, who owned a private bank and was the father of the American composer Marc Blitzstein. Madi's brother Milton ("Mick"), four years younger, and his wife Sabina ("Tibi," a lawyer), lived on the fourth floor of the Leof house. He was a dentist and must

Fig. 1.5 Morris Leof, Charlotte's father, and scion of "322."

have been pretty exceptional because his services were used by all the members of the woodwind section of the Philadelphia Orchestra—people who took their teeth very seriously. The youngest child, Charlotte, born in 1911, was ten years younger than Madi. She went to Girl's High when I first met her, and later to the University of Pennsylvania. She had deep, expressive eyes, dark hair in a page-boy haircut, and was bright, lively, and intelligent. Years later, the novelist Haakon Chevalier described her as "small and birdlike," in his thinly disguised roman à clef, *The Man Who Would be God*. "She was not really pretty," he wrote, "but because of the way she handled herself she was as attractive as if she were."

On weekend evenings, my brother Billy and I often went out with the Leof siblings to the Horn & Hardart's on Broad Street. In Depression days, Horn & Hardart's was a central restaurant with cheap food where they would let you sit around for hours with nothing but a cup of coffee, and it became a kind of bohemian hang-out. We also spent a lot of our weekend time hanging around the living room at 322 and listening to the lively talk, which ranged over everything from politics to art. People would just drop in. Among

Fig. 1.6 Madelin ("Madi") Leof, Carlotte's sister, and husband Sam Blitzstein.

the personalities who visited were playwright Clifford Odets, who gave readings of some of his plays there, and journalist I. F. Stone. Two other habitués of 322 were Jean Roisman and Harry Kurnitz. Jean was a poet who later married the liberal attorney Leonard Boudin, and became the mother of the notorious terrorist Kathy Boudin. As for Kurnitz, he was a man of many talents, from petty thievery to music criticism. If Mrs. Leof left her purse in her bedroom with a twenty dollar bill in it when Kurnitz was in the house, it was more than likely to disappear. He also was in the habit of cutting pages out of books in the public library with a razor blade. He covered the Bach festival at Lehigh for the Philadelphia *Ledger*, signing his articles, "Agnes Dei," and wrote a couple of quite successful novels under the pseudonym "Marco Page." He later married the conductor Leopold Stokowski's daughter. By the time I was at Los Alamos, he had moved out to Hollywood, where he worked on a number of screenplays, including one in the *Thin Man* series and one, a dramatization of his own detective story *Once More, with Feeling*, which still appears on late-night TV. I don't recall how it came to pass, but at some point the famous physicist Wolfgang Pauli and his wife visited 322, were taken by Jean and Harry, and invited them to visit them in Princeton for a weekend. When bedtime arrived, Franca Pauli escorted Jean to a bedroom; a few minutes later, Wolfgang escorted Harry to the same room. The guests were quite astonished to find themselves together—the Pauli's having made an unwarranted assumption—but put up with it rather than embarrass their hosts.

During my junior year at Lehigh, I bought a Model T Ford for $75 and used it for my weekend trips back to Philadelphia. Like all Model T's, you had to get in front and turn a crank to start it. The car had its peculiarities. After a while the clutch bands wore out and, since it had planetary gears, it no longer had a real neutral. When I started the engine with the crank the car would immediately start forward. I had to hold it back with both hands on the radiator, then jump aside and leap into the driver's seat as it went past. In the end, though, I had to learn how to install a new clutch band. Another trick it had was throwing a tire. We'd be going down a road, the ride would suddenly become very bumpy, and the steering wheel would pull over to the side a bit. That meant that a tire had come off the rim, and we'd see it rolling down the road ahead of us. It meant stopping, chasing down the tire and reinstalling it, and starting off again.

Also after a while the car developed a leak in the oil pan gasket, and I never could remember to check the oil level. So I'd be driving along when all of a sudden it would feel as if I were beginning to go up a steep hill. The car would slowly grind to a halt, and smoke would begin to pour out from under the hood. When I opened it, the engine block would be a bright cherry red. I would have to sit for a while and let it cool, then get to the next gas station and add some oil. Those Model T's were wonderful cars; such incidents never seemed to do them a bit of harm.

While I was in college, my uncle Lester got me a number of summer jobs. The first one was at the Philadelphia gasworks, where I was engaged in packaging stuff in the mail-order department. The notable thing about that job was that we were paid in gold. Every two weeks we'd line up in front of a cashier's window in which there was the kind of change dispenser you see in movie houses. We would present our pay slips, the cashier would push some buttons, and gold and silver coins would slide down into a little metal dish.

The next summer I had a job in an oil refinery as a plumber's assistant, learning to be a pipe fitter. The third summer was more interesting. Uncle Lester got me a job on an Atlantic Refining Company tanker. We sailed from Philadelphia to Corpus Christi, Texas, then up to Nova Scotia and the St. Lawrence River and down to Montreal, where we delivered the oil, and then back to Philadelphia. My job was as a fireman, to sit shifts in the engine room and every hour change and clean the burners under the boiler. It was hot work. As soon as one got out from under the ventilator the temperature was about 140 degrees. I did enjoy, though, being a sailor of sorts for a couple of months. The final year, the summer after I graduated, I finally got a job with some real responsibility, with the Sperry Gyroscope Company in Brooklyn. The first day, I was led into a laboratory and turned over to a man whom I was told was my boss. He had a gyroscope on his desk about the size of a fist, run by electric motors and batteries. He started the gyro up and told me, "Now we wait for half an hour until it gets to thermal equilibrium." At the end of half an hour, he set it with its axis of rotation in the horizontal position and aligned a metal pointer next to the end of the axle ("Now wait and see what happens"). After two hours, in the course of which the gyro had drifted a few degrees, he stopped it, made some readjustments, started it up again, realigned the pointer, and told me to wait a couple more hours. This went on for three days—the two of us staring at the gyro, with him occasionally resetting it. On the morning of the third day, some men in blue came, took my boss off to a nuthouse, and I acquired his position.

The apparatus we were playing with was intended to be used in making surveys of oil wells. The gyro would be lowered into a well, and it would stay in a fixed orientation while recording the inclination and direction of the well on paper tape. Its chief purpose, I learned, was in legal suits. It seems it was common practice for drillers to plant their drills near the edge of their property, drill down a ways, then put in a wedge that would shunt the drill off at an angle into a pocket of oil on the neighbor's property. To combat this, the offended party could get a court order to have the well surveyed. In a short time, I got the thing working pretty well and to do what it was supposed to do; it was mainly a matter of simplifying the procedure and putting in sturdier components. I was then sent down to Louisiana to test the thing out in the field. That was a lot of fun, riding around the bayous in the propeller-driven boats and riding the drill pipes up and down in the derricks. When I returned to Brooklyn, the Sperry people were quite enthusiastic and tried very

Fig. 1.7 John H. Van Vleck (1899–1980), who developed the quantum mechanical theory of magnetism, graduated from the University of Wisconsin in 1920 and received his Ph.D. from Harvard University in 1922. He taught at Wisconsin (1928–1934) before moving to Harvard, where he became chairman of the physics

hard to get me to take a job with Sperry. However, during my last year, my teachers had suggested I go on to graduate work, which sounded more interesting to me, and they found me a teaching assistantship at the University of Wisconsin, which I was glad to accept. So I turned down Sperry and left for Madison, Wisconsin.

I was lucky: 1930 was the last year anyone got a teaching assistantship, or any other job, as the Depression deepened. I was lucky in another way too. I was completely innocent of any knowledge of what went on in physics at any American university; I applied where my Lehigh teachers suggested and went to Wisconsin because I was offered a job there. But at Wisconsin I found myself under the tutelage of John Van Vleck, a fine physicist and teacher (who would win the Nobel Prize in 1975) and a kind and considerate man. "Van" had received his Ph.D. from Harvard in 1922 as a student of Edwin C. Kemble, had arrived at Wisconsin in 1928, and was one of the few in his generation who was trained in America, not Europe. During the day, he was usually down in his office, and he made it clear we could go down at any time to talk to him about our problems; and once or twice a week, he would wander by our offices to make sure everybody was doing all right. Another factor that made Wisconsin one of the most lively and interesting schools in the country in those days was the presence of Alexander Meiklejohn's Experimental College, peopled mostly by bright kids from New York.

A teaching assistant got $800 a year and, at Depression prices, we could live on this without too much trouble. A dinner at the Student Union cost thirty-five cents and a midnight hamburger a nickel. Housing in the winter, when temperatures could get to thirty below, was some problem. I roomed with Audley Sharpe, another teaching assistant who doubled as the university's seismologist. Our first apartment had no heat at all: the landlord, who lived below us, preferred to use our monthly $25 for food rather than fuel. We slept there but had to spend our waking hours in a heated office in Sterling Hall—an arrangement that encouraged study. We relaxed occasionally with red wine in a speakeasy in the Italian district—those were Prohibition days. On occasion we would cajole some alcohol from friends in chemistry and add water to make what we called "panther piss." Later we got an apartment with a coal stove in the living room; still, a cup of tea left on the kitchen table in the evening would be frozen solid by morning.

The extreme winter weather had some compensations. There was an occasional auroral display, as well as some elegant hailstones, looking like little tables or stylized mushrooms with a hexagonal top a half-inch across and an eighth-inch thick and a half-inch high hexagonal stem. But the sight that took the prize was an exhibition of sun dogs and other arcane phenomena one winter noon. The sun was about twenty degrees above the horizon. In a ring, every thirty degrees clear around the horizon, were images of the sun as bright as the original. Connecting them was a horizontal band of white. From each a pillar of fire reached the ground and a rainbow rose to the zenith. A pattern of other rainbows circled the sky. The display lasted some time—I couldn't say now whether it was fifteen minutes or half an hour. Later that afternoon, Audley and I went to the library and in a meteorology book found a reproduction of an old engraving which was a dead ringer for what we had seen. The caption said "as reported by a whaling captain in the Antarctic in 1753."

I was thrown into teaching physics lab sections as soon as I arrived, without experience or instruction. My first assignment was a section of "Home Ec" girls. Fortunately for me, we shared the lab with a similar section taught by a young instructor who helped me through the initial traumas. He seemed very wise to me, but it turned out later that this was also the first college class he had ever taught. His name was Leland Haworth. I have a distinct recollection of a girl who brought a lab report to show me, and she had made some funny gaffe or other. In my still-unsophisticated way, I laughed out loud when I read it, which of course hurt her feelings. Lee happened to be standing next to me, and he shook his head sternly and told me not to do that. Later, Lee would use his talents handling people as director of Brookhaven National Laboratory, as a commissioner of the Atomic Energy Commission, and as head of the National Science Foundation.

Van Vleck wasn't actually at Wisconsin when I arrived in the fall of 1930; he was away in Europe that semester, where he attended the Solvay Conference. After his return he gave a course in quantum mechanics, my first exposure to the subject. Van's course treated the principles of quantum mechanics in passing and concentrated on applications to atomic and molecular problems. I also learned a lot, and got to see a good deal of Van, by proofreading the galleys of his classic book, *Electric and Magnetic Susceptibilities*.

No one got a Ph.D. at the end of that first year. If you had a teaching assistantship, you didn't finish your thesis: there were no jobs to be had. So there were no new TAs, and the next year started with the old bunch. Van obligingly gave Advanced Quantum Mechanics and, the following year, Advanced Quantum Mechanics II.

I shared an office for four years with Emanuel ("Manny") Piore. Manny would amuse me with tales of his uncle named Harry Gerguson, better known under the pseudonym Mike Romanoff—Prince Michael Alexandrovitch Dimitry Obolensky Romanoff—about whom the *New Yorker* ran a five-part profile in 1932. Romanoff charmed, amused, and sponged off the very rich. He would stow away, first class, on any trans-Atlantic liner, and was in constant trouble with Immigration. Once, while being detained at Ellis Island, he persuaded the authorities to allow him ashore in the custody of two Immigration officers. He took them to a nightclub, got them drunk, and walked off. He was generous to Manny when flush, and when broke came and stayed with his sister, Manny's mother. Later, Gerguson founded the famous Hollywood restaurant Romanoff's. After World War II, Manny became head of the Office of Naval Research, which financed physics research at the universities, and later was vice president in charge of research at IBM and a director on its board.

Also at Wisconsin was Ragnar Rollefson, an instructor who stayed on to become chairman of the university's Physics Department after the war; Neil Whitelaw, who became a well-known astrophysicist; and Amelia Z. Frank, who received her degree around the time I did. She was the only female physics graduate student; the situation for women in the field wasn't any better in those days than it is now. Amelia later became Eugene Wigner's first wife when he arrived there in 1936, and she died shortly thereafter. There were two Commonwealth Fellows—the British equivalent of Rhodes Scholars—who came to work with Van: a Scotsman named Robert Shlapp who eventually taught in Edinburgh, and Bill Penney, who later was at Los Alamos with the British Mission and went with me to Japan in September 1945 to study the damage at Hiroshima and Nagasaki. He was knighted for his postwar services to the British nuclear enterprise and ended up Lord Penney. Among the undergraduates were Don Kerst, who would stay on to get his Ph.D. from Wisconsin in 1937 (and with whom I would later collaborate at Illinois), and Ray

Herb, who would become one of the world's foremost experts on Van de Graaffs.

Wisconsin's math department was unique in giving courses of direct use to physicists; graduate math courses are normally designed solely for embryo mathematicians. Columbia physicist T. D. Lee tells a story designed to illustrate the relationship between mathematicians and physicists. After a few days in a strange town, a man bundles up his dirty clothes, leaves his hotel, and goes down the street looking for a laundry. He comes to a store with a sign in the window, "Laundry Done Here." He goes in, puts his bundle on the counter, and asks, "When can I have this back?" The clerk says, "We don't do laundry." "But that sign in your window . . . ?" "We paint signs."

Fortunately, things were different at Wisconsin. I took math courses from Rudolph Langer and Warren Weaver. Langer, who later went to Harvard, taught differential equations, partial differential equations, and integral equations—just the things we needed. I took a course in probability from Warren Weaver, who later became the Rockefeller Foundation's director of the division of natural sciences. On the first day, he began the course by saying, "Improbable things happen every day." His course on electricity and magnetism was less satisfying; he tended to become sidetracked by interesting points of mathematics at the expense of the physics. He was trying us out on a textbook, *The Electromagnetic Field*, that he and Max Mason had just published.

We did some self improvement too. I read Dirac's book and a bunch of us went through E. T. Whittaker and G. N. Watson's *A Course of Modern Analysis*, taking turns reporting on a chapter. Bill Penney gave us an informal course in kinetic theory and another in hydrodynamics.

R. H. Fowler, Ernest Rutherford's son-in-law, visited for a while from the Cavendish Laboratory in Cambridge and gave us some lectures on quantum statistical mechanics. I took a course in light from Charles Mendenhall, who was chairman of the department, which was remarkable chiefly for the fact that "Mendy" had never learned vector notation. It would take him half the period just to write out Maxwell's equations in Cartesian coordinates on the blackboard. In 1932 we were excited by the discovery of the neutron, and we held a few seminars on it. But nobody at Wisconsin was seriously working in nuclear physics.

Someone gave Van a mechanical calculating machine that could do arithmetic operations. It was a device about 30" by 18" and built on legs so it stood desk high. It was solid brass with elaborate scrolls and large cranks that you turned to set the numbers, and the word *Millionaire* was stamped in large letters across the front. It was the first calculating machine I ever used. A few years ago, there was an exhibit on the history of calculators at the IBM building on Madison Avenue, and I saw a Millionaire in one of the windows.

In the fall of 1931 the Wisconsin football team played an away game at the University of Pennsylvania. I don't remember who won, but Charlotte Leof, who was attending Penn, used that occasion to send me a note. I hadn't spoken to her in two years, but we began to correspond, and in the summer of 1932, on a trip home, I began seeing her again. Charlotte visited Madison at the Christmas holiday in 1932. During her stay the Van Vlecks invited Charlotte and me to dinner. Van Vleck was a fanatic about trains, and Charlotte was nonplussed and quite unable to respond adequately when, at the dinner table, Van cross-examined her on how often and where her train from Philadelphia had stopped for the locomotive to take on water.

Charlotte and I got married when she graduated from Penn in the spring of 1933. It was a private ceremony. My father took us to City Hall, to a judge who was a friend of his. My sister Alice was the only witness. After the ceremony my father shook hands with the judge and slipped him a bottle of whiskey, an illegal act under Prohibition, but symbolic of the mores of the time. Our honeymoon was a weekend at the Breakers Hotel on the Boardwalk in Atlantic City. For a day, this was quite romantic—then Madi and Jean Roisman, as well as Mick and Tibi of the Leof clan, showed up expecting to freeload in our room. Then we moved into an apartment in Madison for $25 a month. During the following year Charlotte emulated her sister Madi as a freelance journalist, selling stories to nationally syndicated Sunday supplements and to papers like the *Boston Globe* and the *St. Louis Post Dispatch*. She interviewed Frank Lloyd Wright at his estate, Taliesin, where he had his architecture school. Wright was famous for being unfriendly and had set up an intimidating situation: He had her sit down in front of him, and he had collected a dozen students ringed around him. When she asked a question, he would point to a student who was expected to answer.

Van started me on research before the end of my first year. It was a heady time for graduate students in those days; quantum mechanics was brand new and research problems were a dime a dozen. A century of physics which had often been interpreted only in an ad hoc fashion was now amenable to principled theoretical treatment. Almost every subject in molecular and atomic theory had to be reexamined. The problem he gave me was to apply quantum mechanics to the Faraday effect, the rotation of the polarization of a beam of light caused by a magnetic field. I reported on my work at my first APS meeting, the 1931 Thanksgiving meeting, held as always in Chicago. The meeting, located in the University of Chicago physics building, was well attended; about 150 people were there, and everyone gathered in one room to hear all the talks. I remember being embarrassed when I ran over the ten-minute time limit without being finished. The chairman asked me to stop, but, noticing my embarrassment, someone motioned that I be allowed five minutes more, which was granted. I joined the American Physical Society that year. In 1931—a real Depression year—the American Physical Society had only thirteen new members.

Van told me to write up my work as a paper to submit to the *Physical Review*, the flagship of U.S. physics journals. Early the next spring (1932), I presented him the first draft of my Faraday effect paper, which he read, marked up, and returned to me. In the margins next to each equation, he had written a dollar sign and a figure indicating how much it would cost the *Physical Review* to have the equation typeset. Van was on the Board of Editors of the *Physical Review* at the time and conscious of such things. I found I could cut out quite a few, leaving a little algebra for the readers to do on their own.

One day in September 1932, Van entered the quantum mechanics class bearing a stack of reprints—reprints of my just published *Physical Review* article, "The Theory of the Faraday Effect in Molecules" (paper 1), which somehow had been sent to him. He handed them to me proudly while the class applauded and asked me to autograph one for him. Embarrassed and gauche, I asked him what was the usual salutation. He said, "With the compliments of the author." I then wrote down exactly those words, realizing only later the situation called for my writing something warmer like, "with gratitude."

There was one section on band spectra included in the "Faraday Effect" paper which Polykarp Kusch verified thirty years later by running photometer curves on an illustration showing such a spectrum in an old *Physical Review* article.[1] Before electronic publishing the quality of reproduction was good enough so one could actually make such measurements.

I published half a dozen papers (papers 1 through 6) before I got my Ph.D. in 1934. My next one consisted of an application of quantum mechanics to the classical theory of the Kerr effect, the splitting of light into two beams of different velocities induced by an electric field (2). In the course of that work I developed a certain technique for calculating statistical averages at a given temperature which I elucidated in an article published immediately afterwards (3). This was followed by a paper generalizing Dirac's method of calculating the energy levels of many electron systems with the aid of the permutation group (4). When I showed this work to Van, he was surprised because he said he had tried to do the same thing and failed. Papers (5) and (6), "The Solution of Problems Involving Permutation Degeneracy" and "The Energies of Hydrocarbon Molecules," appeared in the *Journal of Chemical Physics*. During the next few years I received a number of invitations to give invited talks at meetings of the American Chemical Society but, well aware of my limited credentials as a chemist, I was always cautious enough to turn them down.

In 1934 I got my first experience in graduate teaching when Van asked me to take over his quantum mechanics class, of which I myself was a member, while he went to Stanford for part of a semester. He gave me a copy of N. F. Mott and H. S. W. Massey's book on scattering theory and told me to teach that topic. It was all new to me. Each day I would go out and tell the class what I had learned the night before.

For four years, I don't remember anyone finishing their Ph.Ds at Wisconsin. No one was going to kick us out. With no jobs to be had we had no incentive to finish, and everyone just stalled. But I took my degree in 1934. I had an embarrassing moment on the final exam. The first question concerned the title of my thesis, and I couldn't remember which of my papers Van had told me to submit.

I took the degree because I had been awarded a National Research Council Fellowship, which carried the princely stipend of $1,200. There were only five given out in theoretical physics in the country. I still have the engraved certificate of award, which is a remarkable document: it is signed by the members of the selection committee, who were the big shots in American physics, mathematics, and chemistry. When Isidor Rabi saw it forty years later he said, "Those guys never let any power out of their own hands." It was signed by A. Flexner, Oswald Veblen, Charles S. Mendenhall, Robert A. Millikan, George D. Birkhoff, E. P. Kohler, Karl T. Compton, F. G. Keyes, F. W. Willard, Gilbert A. Bliss, and Roger Adams.

Van made me go to commencement. In the ensuing sixty years I only went to two others (not counting my sons' graduations from grades one through twelve): one when I got an honorary degree from Lehigh and one when Murray Gell-Mann got one from Columbia and I had to escort him. But in 1996, I went to two more: my son Zach's graduation from Columbia College and, two days later, an occasion at which I received an honorary Doctor of Science degree from the University of Wisconsin.

I was still ignorant of what went on at other American universities. Van suggested I take my fellowship with Wigner at Princeton, and I informed the NRC of my choice. After commencement, Charlotte and I packed all our belongings in a secondhand tan Nash roadster we bought in Madison and started to drive East.

Along the way, we stopped in Ann Arbor to attend the famous physics summer school at the University of Michigan. This summer school had been established a few years earlier by the Dutch physicists Sam Goudsmit and George Uhlenbeck as a way to bring across prominent European physicists, and played an important role in importing the new quantum mechanics into the United States. For me, that summer school was important because we had been pretty isolated in Wisconsin, and here I was introduced to a lot of new people and new ideas, ideas at the forefront of physical theory, in particular the difficulties associated with the field theory aspects of the Dirac equation. But most importantly for me, Ann Arbor was where I first met J. Robert Oppenheimer.

TWO

Berkeley and Pasadena, 1934–1938

In the summer of 1934, Robert Oppenheimer was thirty—tall, thin, and awkward, with bushy dark hair and bright blue eyes. His mind was so quick and his speech so fluent that he dominated nearly every gathering. He was generous and could be very charming. At Ann Arbor that summer I soon became fascinated by him—we got a bit acquainted at a party at which the "drinks" turned out to be lemonade—and decided I wanted to spend my National Research Council fellowship with him. He agreed, provided the NRC approved the change.

Oppenheimer had graduated from Harvard in 1925 and got his Ph.D. at Göttingen in 1927. He spent two years as a postdoc at Leiden and Zurich. At Leiden he picked up the nickname "Opje," which was still used in 1934. By a couple of years later it had become Americanized to "Oppie." After his marriage in 1940, his wife Kitty insisted on "Robert." But he always signed his letters to me "Opje." In 1929 he returned to the United States with a joint appointment at the University of California in Berkeley (where he was an assistant professor of theoretical physics) and the California Institute of Technology in Pasadena (where he was an associate professor). He had established a school of theoretical physics in Berkeley, and it was there I proposed to go.

At the end of the summer session in Ann Arbor, Charlotte and I got in the Nash and continued East for a planned visit to our families. When we reached Philadelphia, I called the NRC office in Washington to ask if I could switch from Princeton to Berkeley. They

Fig. 2.1 J. Robert Oppenheimer (1904–1967) graduated from Harvard University in 1925 and, like many bright young U.S. physicists of the day, went to study in Europe, receiving his Ph.D. from the University of Göttingen in 1927. He was a National Research fellow at Harvard and Caltech in 1927–28. In 1929 he received a joint appointment at the University of California at Berkeley and the California Institute of Technology. He organized the Los Alamos Scientific Laboratory in 1942 and was appointed its director in 1943. He was Chairman of the General Advisory Committee of the U.S. Atomic Energy Commission from 1946 to 1952. He was declared a security risk and barred from government projects in 1954. He was director of the Institute for Advanced Study at Princeton, New Jersey, from 1946 to 1966. (Photo courtesy of the University of California, Berkeley)

said it would be all right if Wigner was agreeable. When I called the Princeton physics department, they told me he was in Europe, had been for months, and probably didn't even know that I had applied to work with him. I called back the NRC, and they allowed that the switch would be all right if someone else in Princeton released me.

One hot summer afternoon, Charlotte and I drove from Philadelphia to Princeton, and of course no one from the Physics Department was in town. A secretary told us that Ed Condon, another eminent physics professor, had a summer place in New Hope, only a few miles away. We found the place easily, and there on the front lawn we found Ed sitting under an apple tree with a drink in his hand. He gave me his blessing, saying that if he were my age he would probably go work with Oppie too.

So in August we turned our old Nash roadster around again and headed back West towards Berkeley. We almost made it; the Nash broke down in Rodeo, California, thirteen miles from Berkeley. We phoned Oppie's office and reached Ed Uehling, another National Research Council fellow who had elected to work with Oppie that year. Ed drove out and collected us and our belongings, and we left the Nash to be repaired.

Ed took us in to the university, which at the beginning of September had already been in session a couple of weeks. He introduced us to Melba Phillips (she changed her name to Melber a few years later), who volunteered to help us find a place to live. Melba had received her Ph.D. the year before, had been Oppie's first doctoral student, and was now an instructor in the Physics Department. Shortly before we arrived, the two had been the embarrassed subjects of a story that achieved wide notoriety. According to the story, the Berkeley police found Melba sound asleep in a car in the Berkeley hills. When they awakened her, she said Oppie had driven her up there, and she had no idea what had become of him. After a search, they found him asleep in his room at the Faculty Club, having apparently walked home and—forgetting all about his girl and his car—gone to bed. The story was picked up by the world press as a classic in the genre of absent-minded-professor tales; Oppie's brother, Frank, saw it in the Cambridge, England, papers. Oppie was still a little defensive about it. His version was that he had told Melba that he was going to walk home and that she should drive the car back, but that she had dozed off and hadn't heard him.

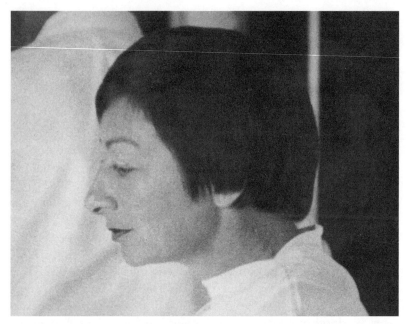

Fig. 2.2   Charlotte Leof Serber. (Photo courtesy of Henry Barnett)

Melba took a few hours off, and we hunted for apartments all around Berkeley without success. Later, we went to a real estate agent who found us a second-floor room in a pleasant enough house. As we were walking down the stairs after agreeing to take it, whom should we encounter but Melba, who turned out to have a room on the first floor. Out of natural concern for her privacy, she hadn't shown us a vacancy in her own rooming house. She was quite embarrassed, but we went on to become good friends.

At the university, the day after arriving in Berkeley, I found that Oppie was indeed the Pied Piper of theoretical physics. Of the five NRC fellows in theoretical physics that year in the whole country, three had chosen Berkeley; the third besides Ed Uehling and myself was Fred Brown. By 1934 Oppie had the liveliest school of theoretical physics in the country. Among his students, past and present, were Harvey Hall, Frank Carlson, Leo Nedelsky, Wendell Furry, and Arnold Nordsieck. Ed McMillan, a postdoc working with Ernest Lawrence, liked to hang out with the theorists; Ed soon became, after Oppie, our best friend. Felix Bloch, at Stanford, was not far away and we often had joint seminars with him and his students.

Fig. 2.3 Group of Oppenheimer students.

A week or two after we arrived, Oppie asked us one evening if we would like to go with him to a movie in Berkeley. That evening became the beginning of a very special rapport between us. I still remember the movie; it was *Night Must Fall*, a thriller starring Robert Montgomery. Somehow, during that evening, the atmosphere changed.

Oppie's students adored him and imitated his speech and mannerisms. Since he had the rare ability to speak in complete English sentences, the result was mostly salutary. Fred Brown was the only one who didn't succumb to Oppie's charm. Once, when Oppie spoke sharply to Fred, Fred's fiery Mexican wife Jovita lit into him and silenced him completely, a unique experience. His students received a cultural education as well as instruction in physics; they learned his tastes in food, art, and music. Bach, Mozart, and Beethoven were acceptable, as were string quartets. Ditto the Impressionists.

Oppie was generous with his students, all of whom were living on shoestring budgets. After seminars Oppie would often take the whole bunch out to a good and expensive restaurant in San Francisco, usually Jack's, an occasion for us who were living through the Depression hand-to-mouth. Once a student, Chiam Richman, ordered roast beef well done, and Oppie turned to him scornfully

Fig. 2.4 Edwin McMillan (1907–1991) received a B.S. in 1928 and his M.S. in 1929 from Caltech, and a Ph.D. in physics from Princeton in 1932. He was a National Research fellow at Berkeley from 1932 to 1934 and then joined the Berkeley Radiation Laboratory staff and the university faculty. He was the co-discoverer of neptunium, and involved in the discovery of carbon-14. He worked at the MIT Radlab and at Los Alamos during World War II. After the war McMillan became interested in particle accelerators and in 1945 discovered (independently of Vladimir I. Veksler) the principle of phase stability, which made synchrotrons possible. McMillan shared the Nobel Prize in Chemistry in 1951 with Glenn Seaborg. After Ernest Lawrence died in 1958, McMillan succeeded him as director of the Lawrence Berkeley Laboratory, from which he retired in 1973. (Photo courtesy of the University of California, Berkeley)

and said, "Why don't you eat fish?" At the end of one dinner at Jack's, after a joint seminar with Stanford, Bloch grew expansive, and leaned over again and picked up the check. He looked at it, blinked, leaned over and put it back down.

Charlotte, Oppie, Ed, and I spent many evenings in San Francisco. The Bay Bridge was still under construction, and to travel across the bay between Berkeley and San Francisco one had to take the ferry. The ferry terminal at the San Francisco end was the locus of low-down bars, at which one waited for the next ferry, and we usually saw a fight or two and missed a couple of ferries before getting home. At a Mexican bar one night, where we were drinking tequila with a water chaser, the waitress passed around a dish of hot peppers, and the one I got ignited my mouth. I grabbed up a full glass of water, drank it straight down, and not until I got to the bottom did I realize it wasn't water but straight gin. Someone had made a mistake with the chaser. Oppie's Packard roadster, named Geruda after the Sanskrit messenger of the gods, had a rumble seat, and Oppie and Charlotte put me in it so I would get plenty of air on the return trip.

At that time, a bitter longshoreman's strike was taking place in San Francisco, and feelings in the community ran high. Leonard Loeb, one of the physics professors, even tried to get the Cal football players to act as strikebreakers. One day, Oppie told us that there was going to be a big rally of the longshoreman's union in San Francisco that night, and that a friend of his, Estelle Caen, had asked him if he would like to attend. He wondered if we would like to go along. We did, and met Estelle there for the first time. She was the sister of Herb Caen, a columnist from the *San Francisco Chronicle*, who spent some sixty years at that paper and eventually won a Pulitzer Prize in 1996, the year before he died. The meeting was held in a huge auditorium, something like the old Madison Square Garden. We were sitting up high in a balcony, and by the end were caught up in the enthusiasm of the strikers, shouting with them, "Strike! Strike! Strike!" Later, we met Harry Bridges, the leader of the longshoreman's union, in Estelle's apartment.

As far as work goes, I immediately got involved with the nuclear physics going on in the Berkeley Radiation Lab. My only other contact with the subject had taken place during the previous year in Wisconsin, when I read Heisenberg's 1932 article on nuclear physics,[1] which can be described as the Old Testament of the subject.

Milt White was doing an experiment to try to detect proton-proton (*p-p*) forces, scattering 0.7 MeV protons from the 27" cyclotron from hydrogen in a cloud chamber. I derived the modification of the Mott scattering formula which a *p-p* force would produce, found a ^{1}S phase shift which I guessed was the best fit I could get to Milt's rough results, and calculated the well depth required to give that phase shift. For a square well of range equal to the classical electron radius, which was a conventional choice in those days, the depth was 12.7 MeV—within a few percent of the true answer, 11.8 MeV. I remember meeting Milt at the door of the Radiation Lab and his asking me for the number, which I gave him from memory. So it may be my fault that the numbers got transposed and the depth was given as 17.2 MeV in Milt's paper.[2] Milt's experiment antedated the first measurement of the low-energy neutron-proton (*n-p*) cross section, by Dunning, Pegram, Fink, and Mitchell, which was published in the subsequent issue of the *Physical Review*,[3] a fact I discovered years later when I wondered why we hadn't noticed the equality of the forces.

Ed Uehling was calculating the effect of vacuum polarization on the energy of atoms. Vacuum polarization was one of the surprising consequences of the Dirac equation. According to this theory, an electromagnetic field can momentarily produce an electron-positron pair out of vacuum. Since the field pushed electrons and positrons in opposite directions, this produced a polarization of the vacuum just like the polarization of a dialectric medium, and as a consequence, the Maxwell field equations needed to be modified. A point charge, for instance, no longer produced a pure Coulomb field that fell off with the inverse square of the distance. At distances less than the Compton wavelength of the electron, 4.8×10^{11} centimeters, there would be appreciable corrections to this law. There was a hitch, though; when one calculated the dialectric constant, it turned out to be infinite. Infinities were common in the electrodynamics of the 1930s. Willis Lamb, as a graduate student, was once confronted with a perfectly finite integral and gave an order-of-magnitude estimate of it. It didn't occur to him that an integral arising in a field theory problem could actually be evaluated. And Felix Bloch once remarked, "Just because a quantity is infinite doesn't mean that it's zero!"

To solve the vacuum polarization problem, Dirac proposed a

scheme for subtracting off infinite terms, so that one was left with a finite result. Heisenberg took up Dirac's proposal and calculated in detail the terms that had to be subtracted, but both of them concentrated on the infinite terms and hadn't fully evaluated the finite terms that were left. This was a problem that Ed Uehling was working on—calculating the vacuum polarization for an electrostatic field. In the uncorrected Dirac theory, the ^{2}S and ^{2}P states of the hydrogen atom would have the same energy. Since the ^{2}S electron spends more time closer to the nucleus, the vacuum polarization would have a bigger effect on this state, and one would expect a separation in energy of the two levels. A team of spectroscopists at Caltech claimed that there was already evidence for such a separation, to the amount of .03 wave numbers, though the evidence was not universally accepted.[4] Oppie thought that perhaps vacuum polarization was responsible, and set Ed to calculating the expected effect. But Ed found that the line shift produced by vacuum polarization was thirty times smaller than that suspected experimentally.

At the beginning of the Thanksgiving vacation Oppie gave me Heisenberg's paper on positron theory[5] and sent Charlotte and me in our now repaired car on a week's trip into the Mojave desert, during which I was supposed to read it. On returning I generalized Ed's calculation of the vacuum polarization corrections to an electrostatic field to get the corrections for an arbitrary electromagnetic field (the e^2 correction to the photon propagator, in modern terminology). I also verified that the effects vanished outside the light cone, as required by causality. My paper, my first from Berkeley, was entitled "Linear Modifications in the Maxwell Field Equations" (7) and was sent to the *Physical Review* in April 1935 and published back-to-back with Ed's paper in July.

As Oppie held a joint appointment at both Caltech and Berkeley, while Berkeley was in session he would spend a day a month in Pasadena. Berkeley ran on an atypical academic calendar—starting in the middle of August and ending before April—and Oppie could spend the last quarter of Caltech's normal academic year in Pasadena. Some of his students would traipse down with him. We didn't have more possessions than could be carried in our cars, so we would give up our Berkeley apartments and rent Pasadena garden cottages, which could be had for $25 a month—even cheaper than the $40 common in Berkeley.

At Caltech "Oppie" metamorphosed into "Robert." We consorted with Charlie Lauritsen's gang: Willy Fowler, Tommy Bonner, Louis Delsasso ("Del"), and Richard Crane. Charlie's son, Tommy, was an undergraduate. Tommy made pocket money by manufacturing Lauritsen electroscopes and selling them to other labs. One day he was showing around a check from the Cavendish Lab for $25.87, signed simply "Rutherford." Tommy said he was tempted to have it framed, and would have if the eighty-seven cents for postage hadn't represented a cash outlay. There was also Stuart Harrison, an English radiologist who treated cancer patients at Charlie's million-volt X-ray tube. The alchemist's dream of those days was to find the cure for cancer, and every lab that had a high-voltage accelerator had a cancer research program.

The X-ray tube was in the Kellogg Radiation Lab in a cavernous room about fifty feet high. It was a structure fourteen feet tall on top of a flat-roofed shack that provided lab space. That accelerator had more pizazz than any of the others I've been associated with. At Kellogg, a metal ball, suspended from the ceiling and swung by a rope towards the high-voltage end, drew very authentic lightning flashes, accompanied by real thunder. Once Willy Fowler and Delsasso somehow enticed Charlotte to stand on top of the lab shack and turned on enough voltage to make her hair stand on end.

That was the Kellogg way. I remember one party, a housewarming for Stuart, who moved from the Athenaeum to a second-story apartment. There were real goldfish in the toilet and ducks in the bathtub. At one point Charlie Lauritsen, on the back porch, picked up a large round bath mat (which was soaked because, naturally, someone had forgotten to turn off the water while filling the bathtub for the ducks), and saying, "Here's how I used to net fish in the Everglades," let fly. He scored a perfect hit twenty feet below on Maisie, the X-ray nurse, whom Delsasso was chasing around the backyard. Charlie really had been a fisherman in the Everglades when he first came to this country from Denmark; he said he had looked at a U.S. map for a place with a warm climate where life would be cheap and easy. On the map the Everglades seemed a perfect choice.

After the evening journal club that met once a week at Kellogg there would be a party for the participants at Charlie's. It spread between the living room and the back garden, and mainly involved drinking and singing. Besides Danish and German favorites there were American folk songs, like "The West Virginny Mountains" and "Her Name Was Lil." Ed McMillan, whose family lived in Pasadena, came when he was home. Another favorite activity was a Mexican dinner in Olvera Street.

Richard and Ruth Tolman were also good friends. Richard, an expert in relativity and statistical mechanics, was dean of Caltech's Graduate School. Another member of the Caltech staff was the Polish-born mathematical physicist Paul Epstein, who dressed and acted just like Emil Jannings in the early part of *The Blue Angel*. One day, at lunch at the Athenaeum, he regaled his tablemates with the story of his expedition the previous Sunday to the beach, where he spent half an hour watching the sea lions playing on the rocks. He finally decided to count them: there were two.

Willy, Del, and Charlie were measuring the upper-limit energies of the positrons emitted by the mirror nuclei, from ^{9}B to ^{17}F. This was Willie's dissertation. They noted that the energy differences between mirror pairs could be ascribed to Coulomb energy differences. Since the mirror pairs differed by interchange of proton and neutron numbers, they concluded that n-n and p-p forces were equal. My recollection is that this was an independent discovery, though Young's papers,[6] suggesting charge independence on the basis of the systematics of nuclear stability, was published before theirs.[7] Willie, considering the nucleus as a uniformly charged sphere, said the Coulomb energy was $(3/5)(Ze)^2/R$. As a student of Van Vleck's, I knew all about calculating Coulomb energies and worked it out, including exchange terms, for a shell model and for a Fermi gas. But Charlie and Oppie said the accuracy of the data did not warrant such high-falutin' theory, and, much to Willie's annoyance, in the published paper simply gave a charge radius—a model independent statement. The paper was entitled "Radioactive Elements of Low Atomic Number" and, perhaps because its significance was not advertised in the title, nobody noticed it. A year later, in 1937, Wigner rediscovered the mirror nucleus symmetry and Hans Bethe discussed the pair ^{3}He–^{3}H.[8]

In early June of 1935, Dirac came by Pasadena on an around-the-world trip. One day, Oppie brought him into a room where Arn Nordsieck and I were working, introduced him, and told him that the problems we were working on were sequels to Dirac's own work. Oppie left, Dirac sat down, and Arn and I gave a fifteen-minute résumé of what we had been doing. When we finished, we turned to Dirac and braced for his comments. He said, "Where is the nearest post office?" I said, "We'll take you there, and on the way we can discuss our problem." Dirac said, "I can't do two things at once."

A few days later, Dirac left on a boat for Japan. Oppie escorted him to the pier. On the way, they passed a bookstore, and Oppie offered to buy some books for him to read on the long trip. Dirac said, "I never read. It interferes with thought."

When the term ended at Caltech that summer of 1935, Oppie invited Charlotte and me to the ranch in New Mexico. We drove out on Route 66, which in those days was a two-lane road, oiled in Arizona, dirt in New Mexico. These were still Depression days, and at every gas station along the way one could buy Navajo rugs for three to five dollars—rugs the Navajos had traded for gas—rugs worth thousands today. When we arrived, Ed McMillan, Melba Phillips, and Oppie's brother Frank were already there. Our introduction to the ranch was typical Oppenheimer style; he said another guest was coming and the ranch was getting a little crowded so why didn't we take two horses and ride to Taos, pointing north, eighty miles and three days away across the mountains. The land north of the ranch was a federal wilderness area where it was forbidden to build any kind of camp or shelter and, as we soon discovered, one would ride all day without seeing another soul. In addition, we had to cross Jicoria Pass at 12,500 feet. Moreover, I'd never been on a horse before in my life. Charlotte had ridden a little in the East, English saddle. Well, we started out. In the late afternoon of the first day, following Oppie's written directions, we dropped down from the high country to find a little Mexican-American village where we could spend the night. I was really suffering by then, feet bruised by the stirrups, muscles aching, and about a square foot of skin rubbed off. Even so, we were more concerned about our horses Blue and Cumbres than about ourselves. We had been given strict instructions about their care: not to overtire them, to loosen the cinch every three hours and let them rest, to dismount and lead them if the going got rough. For the last two miles downhill the poor beasts had their heads between their knees and could hardly put one foot in front of another. But we finally reached the house where, Oppie told us, we would be put up for the night. The owners put our horses in their corral, fed us dinner, and afterwards we were resting on their front porch when two horses came gaily trotting down the road. It couldn't be! Our horses, escaped from the corral and miraculously restored to health and vigor? It was. That began our education in the wiles of horses and the equine sense of humor. Our hosts got out a pickup truck and we rounded Blue and Cumbres up a mile down the road.

The second night we camped out, and we reached Taos on the third day, on schedule. Oppie had told us to put up at an inn in Ranchos de Taos, so we took our horses to the inn's corral. The Oppenheimers were always very solicitous about the horses and concerned that they didn't bear too much weight, so we'd started out in true ranch style with a change of socks and underwear and a toothbrush each, a box of chocolate graham crackers, and a pint of whiskey and lots of oats. The only luggage we had was the horse's feed bags, so, hot, dirty, and tired, we carried our underwear, socks, and toothbrush into the hotel lobby in them. The hotel owner, a Greek, intercepted us in the lobby, but when he heard we came from Oppenheimer, greeted us warmly and personally escorted us to our room. A shower was the first thing indicated, but I couldn't get my boots off. I asked for help and Charlotte, who had just undressed, took up, nude, the classic boot-removing position, her back to me, one boot held between her legs and the other on her rump prepared to push. At this moment we became aware that the large picture window of our ground-floor room faced a low fence, a row of hitching posts, and a row of Indians staring in at us with traditionally unexpressive faces. After our showers we reentered the lobby just in time to hear the proprietor tell some other guests in a loud stage whisper, "We have some very *clessy* guests. We have the assistant to the assistant to Millikan!"

Years later, in 1948, as an after-dinner speaker at Robert Millikan's eightieth birthday party at Caltech, Oppie told this story to illustrate Millikan's standing in the American consciousness; Einstein's was the only other scientist's name that would be recognized, let alone revered.

The next day we started the return trip. We were to meet Oppie, Frank, and Ed at Truchas Town, another small Spanish-American village. Late on the second afternoon, about five miles from Truchas, we were riding through a pine forest, when suddenly Charlotte slipped off her horse and fell to the ground. She was lying on her side, and when I dismounted and knelt over her, I saw what seemed to be a bloody straw sticking out of her cheek. I tried to pull it out, and when I touched it discovered it was a fine fountain of blood about a foot high. A pine needle had pierced an artery, and we were unable to staunch the flow of blood. Charlotte held a rag over her cheek, and, there being nothing else we could do, I helped her back into the saddle and we continued towards Truchas. When we broke out of the woods near the town, we were at the edge of a large meadow. A quarter mile away at the other edge, we saw Frank, Oppie, and Ed. I tried to keep Charlotte to a walk, but she put her horse to a gallop, and when she reached them fell off again. They gazed in amazement; she was lying on the ground with her jacket completely covered with blood. Their reactions were typical of each. Oppie stood there, looking concerned and solicitous. Ed kneeled over and pulled away the cloth over her face, saying, "Let me see! Let me see!" Frank turned his back and said, "Come look at the beautiful sunset."

The ranch was a quarter-section area (one-half mile on each side) on the side of Grass Mountain, above the Pecos River. It stretched between an altitude of 9,000 to 10,000 feet. There was a log cabin, and below it a corral for the horses. For the first few days there, any physical task left one gasping for breath. The first floor of the ranch house had a large living room and kitchen. In the living room was a large fireplace with a couch in front of it, and an Indian rug on the floor. That was it. The kitchen had a sink and a big wood stove. There was no bathroom, but a sort of outhouse at the end of the outside covered porch. Frank, who was stronger and handier than Oppie, had put in a pipe to a spring above the house, so there was running water in the kitchen and toilet. There was an unfinished upstairs that could be used as a bedroom, but it never was before Oppie was married. We all slept on cots on the porch.

There were half a dozen horses in the corral. Oppie's horse Crisis, Frank's horse Pronto, Charlotte's horse Blue, my horse Cumbres, and a couple of others. The business of the ranch was riding. We rode in the wilderness area, in the pine and birch forest, the high grassy and flowered meadows, and along the ridges of the Sangre de Cristo Mountains, as far north as the Truchas Peaks, rising 13,500 feet. We would go for anything from a day to a week. The trips were as sparsely provisioned as on our journey to Taos. Ruth Valentine, a Pasadena psychologist, once gave an apocryphal account of a typical

Fig. 2.5 On the porch at the Oppenheimer ranch.

Oppenheimer excursion: It is midnight, and we are riding along a mountain ridge in a cold downpour, with lightning striking all around us. We come to a fork in the trail, and Oppie says, "That way it's seven miles home, but this way it's only a little longer, and it's much more beautiful!"

Talk of physics was forbidden at the ranch, except on the rare occasions when we had visitors. George Gamow came by once, and shocked me by telling us that the illustration on the cover of his most recent book, which purported to show the picture which led to the discovery of radioactivity by Becquerel, an accidently exposed film with the shadow of a key on it, was in fact faked and produced by Gamow himself, waving a flashlight over a key lying on a piece of film. George Placzek visited once. The Weisskopfs also visited, and once Oppie ran into Bethe hiking in the neighborhood and brought him up to the ranch. Walter Elsasser was a less fortunate visitor. The night he arrived, Oppie served one of his chili concoctions, which had been sitting on the stove for about a week, getting hotter and hotter. The rest of us already had callused mouths. When Elsasser tried a spoonful, tears came to his eyes. He bravely swallowed a mouthful, then turned to Oppie and said, "Is it right?" Elsasser had never ridden. Oppie persuaded him to try a short trip. He put him on Blue, Charlotte's horse. Blue took a couple of tentative steps, sensed that he was in command, and galloped off down the hill to the corral, opened the gate with his nose—which was a trick we had no idea he knew how to do—entered, and brushed Elsasser off under the low roof of an open shed.

We spent time at the ranch every summer from 1935 to 1941, and the visits I mentioned were spread out over this time. Ed McMillan and Melba were at the ranch in 1935; no one else from Berkeley was there in subsequent years.

Back in Berkeley in mid-August we resumed our normal occupations. Charlotte was working again as a freelance journalist. I remember one day a young author, visiting our apartment to be interviewed, was William Saroyan, whose book, *The Daring Young Man on the Flying Trapeze* had just been published. He was still rather diffident, a trait he outgrew later in his career. Charlotte sold the story to the *Boston Globe*, the first large-scale publicity Saroyan received. Charlotte was also active in the Berkeley chapter of the League of Women Voters.

I began to study a new paper of Heisenberg's, which was the first to consider a complete quantum field theory, with both electromagnetic and charged fields quantized. Earlier work had treated one or the other as classically given. Heisenberg proposed to eliminate the divergences by a subtraction method like that Dirac had used for vacuum polarization, but gave it up when it failed to eliminate the electron self-energy divergence. I discovered an error in Heisenberg's calculation: the self-energy singularity was indeed eliminated and one could calculate the electron propagator to order e^2. However I found that the subtraction scheme would fail in higher orders (8). According to Abraham Pais, in *Inward Bound*, this paper is notable for introducing the term *renormalization* into the lexicon of physics.[9]

In June 1936 the American Physical Society held a meeting in Seattle. No one paid our traveling expenses to meetings in those days, and the only meetings we attended were APS meetings on the West Coast. We drove up with Oppie from Pasadena and met the rest of the Lauritsen contingent there. The trip was a fascinating introduction to the big trees and rugged coast of the Pacific Northwest. I gave my first nuclear physics paper (9) at that meeting, interesting now in revealing the foibles of the time. First, the "saturation" of the forces leading to a nuclear binding energy proportional to A was to be explained, not in the obvious way by a repulsive core, but by exchange forces. Second, the nuclear forces were explained as a consequence of the weak interaction. Not much was known of the form of the beta-interaction. Fermi's vector coupling did not explain spin changes; distorted experimental energy spectra seemed to require derivative couplings. I tried to turn the argument around and determine the weak coupling so it would give properly saturating nuclear forces. I don't remember what the conclusion was: the published abstract of the talk leaves it as a surprise for the audience.

Another paper (10) at the Seattle meeting, by Oppie and myself, presented an estimate of nuclear-level densities based on a Fermi gas model, a calculation inspired by the appearance earlier that year of the Breit-Wigner formula and of Bohr's description of a nuclear reaction as formation of a compound nucleus and division of its excitation energy among many particles. It was a hot subject; our abstract and a similar consideration by Bethe[10] appeared in the same issue of the *Physical Review*.

Oppie also gave his first report on electron-positron showers. At that time the primary cosmic rays were supposed to be high-energy electrons. A high-energy electron going through matter produces photons, which in turn produce electron-positron pairs, which in turn produce more photons, so you wind up with showers of electrons and positrons. Oppie said shower theory accounted well for the cosmic ray showers and bursts above ground level, though the source of deep underground showers was unexplained. Cosmic ray problems were much on our mind since both Caltech and Berkeley were active in the field, R. B. Brode at Berkeley and Millikan, Carl Anderson, I. S. Bowen, and H. V. Neher at Caltech. Our relations with the Caltech cosmic ray people were not as informal as with the nuclear physicists, and communication was mainly through Oppie. Carl Anderson and Seth Neddermeyer were the only ones I had much contact with.

After the meeting I stayed and taught summer school at the University of Washington in Seattle. Fred Schmidt, a friend and a professor at the university, rented us a house in a new real estate development. Our house was on the flats; the more expensive houses were built on scooped-out areas in the hills behind us. One day, friends of Fred's invited us out sailing on Puget Sound, and we had a pleasant afternoon until a real downpour started. When we returned home, we found mudslides in the hills and floods on the flats. Our poor dog, who had been locked in the basement, was shivering on the top step of the basement stairs: mud and water filled the basement right up to that level.

My NRC fellowship had expired at the end of the spring semester of 1936. Oppie managed to get me an appointment in the Berkeley physics department as his research assistant. He had surprising difficulty in getting the appointment through. Raymond Birge, head of the department, reluctantly provided a salary of $1,200 a year, and Oppie persuaded Ernest Lawrence to come up with another $400.

By the time of my arrival in Berkeley, Oppie's course in quantum mechanics was well established. Oppie was quick, impatient, and had a sharp tongue. In the earliest days of his teaching he was reputed to have terrorized the students. Now, after five years of experience, he had mellowed—if his earlier students were to be believed. His course was an inspirational as well as an educational achievement. He transmitted to his students a feeling of the beauty of the logical structure of physics and an excitement in the development of science. Almost everyone listened to the course more than once, and Oppie occasionally had difficulty in preventing students from coming a third time. One Russian woman attempted to come a fourth time, and defeated Oppie's efforts to dissuade her by going on a hunger strike. The basic logic of Oppenheimer's course in quantum mechanics derived from Pauli's article in the *Handbuch der Physik*. Its graduates, Leonard Schiff in particular, carried it, each in his own version, to many campuses. Part of my duties as Oppie's research assistant was to take over the teaching of this class at times when he was out of town—in Pasadena, for instance.

Oppie's way of working with his research students was also original. His group would consist of eight or ten graduate students and about a half dozen postdoctoral fellows. He would meet the group once a day in his office. A little before the appointed time its members would straggle in and dispose themselves on the tables and about the walls. Oppie would come in and discuss with one after another the status of the student's research problem, while the others listened and offered comments. All were exposed to a broad range of topics. Oppenheimer was interested in everything, and one subject after another was introduced and coexisted with all the others. In an afternoon we might discuss electrodynamics, cosmic rays, astrophysics, and nuclear physics. As Oppie finished with each student, he would advise him how to continue. Finally, he would leave. Most of the crew would stay, and my job began: to explain to each what Oppie had told him to do. They were much more willing to display lack of understanding to me than they were to Oppie.

Oppie and I often worked together at night in the apartment on Shasta Road to which he had moved about the time we arrived in Berkeley. It was a typical Berkeley hill house, with a splendid view of San Francisco and the Golden Gate. Oppie's apartment was on the lower level of the house. His landlady, Mary Ellen Washburn, whose husband taught economics at Berkeley, lived on the upper level, which was street-level at the front of the house and thirty feet above ground at the rear.

Oppie had a medium-sized living room, paneled in dark wood, with the ubiquitous Berkeley fireplace and a couch and a desk. There was a small kitchen, bathroom, and bedroom. The windows were always wide open, summer and winter, a reminder of the lung problem that had driven Oppie to New Mexico in the first place. One suffered in the winter; it was cold. In addition to the fireplace, there was a gas floor heater, covered with an iron grill. When we were too lazy to light a fire, I often stood on the grill while Oppie paced up and down. I stood on it until the smell of burning leather warned me that my shoes were suffering.

Before starting work, we would sometimes go to Oakland for a Mexican dinner, and when we got discouraged we would take in a movie. Charlotte, who usually came along, would be somewhere in the background while we worked, reading.

Part of Oppie's social life was entwined with his group of students. We had many parties, some of them at Oppie's apartment, at which we drank, danced, ate, and of course talked physics. When Oppie supplied the food, the novices suffered from the hot chili Oppie served, which social example required them to eat. But Oppie had another social life with friends from other areas of the university, and there a was whole other part which had to do with girlfriends— such as Estelle Caen, Jean Tatlock, and Sandra Dire Bennett—whom his students never met. I never saw Jean with Oppie, for instance, and only knew her because she was a friend of Mary Ellen's.

By this time we had, however, become good friends with Oppie's brother Frank, a graduate student in physics at Caltech. During our visit to Pasadena in 1936, he built a phonograph for us, an early example of home electronics. He finished around midnight one evening, and turned it up loud to test it out. He must have woken up the Caltech campus as far away as the Athenaeum. That October he married Jackie—Jacquenette Quann—and the next spring, Charlotte and I shared a house with them.

In 1937, a year after Oppie's exposition of shower theory at Seattle and a month after the discovery of intermediate mass particles (muons) in the penetrating component of cosmic rays by Anderson and S. H. Neddermeyer and J. C. Street and E. C. Stevenson,[11] Oppie and I wrote a letter to the *Physical Review* (12) suggesting that this might be the particle postulated by Hideki Yukawa to explain the *n-p* exchange force.[12] Although Yukawa's paper appeared in 1935, we had never seen a reference to it and knew of it only because Yukawa had sent Oppie a reprint. One purpose of our letter was to bring it to attention. However, we complained that we did not see how, without "extreme artificiality," Yukawa's theory could reconcile the equality of like and unlike forces with their saturation character, or explain the magnetic moments of proton and neutron. At the time it was believed that the primary cosmic rays were electrons and positrons, and we suggested that the mesons were secondaries, photoproduced in nuclei as well as pair-produced. We said that knock-on electrons accounted for the showers produced under heavy absorbers at sea level. The contradiction between a strong nuclear interaction of the mesons and their great penetrating power was not mentioned.

Yukawa had suggested that it was the meson, not the nucleon, that was weakly coupled to electron and neutrino and that beta decay took place through virtual mesons. Our original intention in writing the paper was to point out that, if Yukawa were right, the meson would have a microsecond lifetime so their absorption in air would be greater than in solid materials. But Millikan objected when Oppie broached this idea, saying his own absorption measurements in Lake Arrowhead proved no such effect existed. Oppie said we should rewrite the paper in a form that would not antagonize Millikan. Uncharacteristically, he left the rewriting to me, but no matter how I put it he was not satisfied. After he had rejected my fourth draft, I said in exasperation, "Let's cut it completely." Oppie blinked at me and said, "Do you really think so?" Out it came. It left the paper with an up-in-the-air feeling. A year later Heisenberg and H. Euler made the same point,[13] to Bruno Rossi's enlightenment; he was observing the effect in his experiments in Eritrea.[14]

Early in 1937 Bohr visited Berkeley and gave three public lectures. During the first I was sitting next to Birge, the chairman of the Physics Department, who was the official scribe. Bohr was attached to a microphone, in whose cable he occasionally became entwined and had to be unraveled, but Birge was the backup. Birge scribbled assiduously during the prescribed hour and exactly at its end snapped his notebook shut. Bohr went on talking. After a few minutes Birge opened his notebook and made an entry. I looked over his shoulder and saw "If you will give me a few minutes more." Notebook closed. Five minutes later, "If you will indulge me a moment longer." Bohr went on talking for another half hour—but Birge's notebook remained shut.

Bohr's research assistant, Fritz Kalckar, came with him and stayed on with us for several months. We became good friends. I remember driving down the coast with him on our annual trek to Pasadena and then a week's trip into the Borego desert. While we were still in Berkeley, Fritz and Oppie and I wrote a paper, "Note on Nuclear Photoeffect at High Energies" (13), in which we attempted, not very successfully, to understand nuclear reaction theory when levels overlap. And then from Pasadena, back under Charlie Lauritsen's influence, we wrote "Note on Resonances in Transmutations of Light Nuclei" (14). Our point was that, in the reactions $p + {}^{11}B$ and $p + {}^{19}F$ where energetic alphas could be emitted, both very wide and very narrow levels were observed, and some strong selection rule must be operating to inhibit alpha emission from the narrow levels. We suggested that the total spin and angular momentum were both very nearly constants of the motion, providing a selection rule against transitions from triplet to singlet spin states. As of the spring of 1937 the tensor force had not been thought of.

But not for long. At the APS meeting at Stanford in December 1937 I gave a paper, "On the Dynaton Theory of Nuclear Forces" (15), which was the first effort to get realistic *n-p* forces from a Yukawa-like theory. I don't know where "dynaton" came from, evidently a local invention. I used both scalar and vector mesons (the first application of Alexandre Proca's vector meson theory),[15] the former to give central forces, the latter, with a tensor coupling to the nucleons, to give spin-dependent and tensor forces which I said accounted for both the observed spin dependence and for the anomalous magnetic moments of the nuclei. But since I considered only charged mesons, I couldn't account for charge independence. Although, in the next few months, Yukawa, Sakata, and Taketani,[16] Kemmer,[17] and also Bhabha[18] independently derived tensor forces (and introduced neutral mesons to get charge independence), it didn't occur to any of us to think of the effect on the deuteron. It wasn't until 1939 that Rabi and company[19] discovered the deuteron quadrupole moment, and Rabi was quite unaware that tensor forces had been predicted.

At the same Stanford APS meeting Willis Lamb and I gave a paper on the theory of neutron-deuteron reactions (16).

Charlie Lauritsen's gang had come up from Pasadena for the meeting, and we all stayed in the same motel in Palo Alto. One night we had a little party. After a while the party spilled out into the court-yard of the motel, along with all the furnishings from a couple of the rooms—beds, bureaus, chairs, tables, everything—to make room for a dance floor. We were a little noisy, and the proprietors of the motel didn't appreciate our antics. The next morning, we were requested to leave.

At this time, the Spanish Civil war was in progress, and in Berkeley we were all emotionally involved on the side of the loyal-ists—even Oppie, who never before had any interest in politics or social causes. I think his change was largely the influence of his Aunt Hedwig who, together with her son Alfred and his family, had recently escaped from Nazi Germany and settled close by in Oakland. Jean Tatlock, the daughter of an English professor at Berkeley, who was studying psychiatry at Stanford Medical School, was undoubtably another influence. She was a beautiful and very intelligent woman who was fairly difficult to get along with and sub-ject to fits of deep depression; she and Oppie were having an intense on-again, off-again affair.

Our mail from home told us that Charlotte's father, Dr. Leof, was head of the Philadelphia chapter of the Medical Aid Committee for the Spanish loyalists. There was a tale of a meeting at the Leof home, at which Dr. Leof was trying to enlist a number of doctors in the cause. When one of them suggested that they send medical supplies to both sides, Charlotte's mother, Jenny, who was sitting on the stairs listening, burst out, "Medicine we should send the Fascists? Poison we should send them!" With Oppie's encouragement, Charlotte organized a chapter of the Medical Aid Committee in Berkeley and became secretary of the chapter. Her activities consisted of arranging cocktail parties at which money was collected for medical aid.

The teacher's union had recently been established on the Berkeley campus, and Oppie and a number of his students including myself joined. The issue the teacher's union was pushing at the time was more pay for teaching assistants—a worthy cause.

I think it was at the January 1938 New York meeting of the APS that Oppie had a conversation with Gregory Breit, who pointed out that the charge independence of the strong forces would lead to isotopic spin being nearly conserved for the lighter nuclei, with consequences for such things as selection rules. This led us to reconsider our 1937 paper with Kalckar. In a paper entitled "Note on the Boron Plus Proton Reaction" (17), Oppie and I discussed a striking case. The 0.16 MeV resonance level observed in the bombardment of ^{11}B by protons is a level of the combined nucleus, ^{12}C, with an excitation energy of 16 MeV. Long-range alphas are emitted leaving ^8Be in its ground state and also 16 MeV gammas corresponding to a transition to the ground state of ^{12}C, with a yield 10 percent that of the alphas. What inhibited the emission of long-range alphas to the point that gamma emission could compete? We now said that the 16 MeV state was the first $T = 1$ state, with $J = 2$, allowing gamma emission to the ground state of ^{12}C, $T = 0$, $J = 0$, but forbidding the alpha decay, since both alpha and ^8Be have $T = 0$. We overestimated the validity of the T selection rule, saying it might hold to a part in 20,000 (the square of the fine structure constant). It is now known that the width of the resonance level is 6.5 KeV, most of which is due to decays emitting a shorter-range alpha and leaving the ^8Be in an exited state with $T = 0$, $J = 2$ at 2.9 MeV. The effectiveness of the T selection rule can be seen by comparing this with the decay of a lower-energy resonance level of ^{12}C for which the alpha emission is allowed; the $T = 0$, $J = 0$ state at 10.3 MeV decays to the ground state of ^8Be (also $T = 0$, $J = 0$) plus alpha, with a width of 3 MeV, five hundred times wider.

In addition to giving the first example of an isotopic spin selection rule, we also gave the first example of an analogue state illustrating the full isotopic spin symmetry. The 16 MeV state of ^{12}C is the $M_T =$

0 level of the T = 1 triplet. We said the M_T = -1 level would be the ground state of ^{12}B. This turned out to be not quite correct; the analogue state is actually the first excited state at 0.95 MeV. Following the example set by Fowler, Delsasso, and Lauritsen (note 7), we chose a title that concealed the paper's significance. It went unnoticed, and analogue states were not rediscovered until 1952, when the subject was developed by Robert Adair.

My next paper (18) was on cosmic rays. A graduate student, Hartland Snyder, who was the best mathematician of our Berkeley group, had just given an improved version of Oppenheimer and Carlson's shower theory. Hartland, who came from Utah, was reputedly an ex-truck driver. His speech and manners were a little rougher than our upper-middle-class standard. We called him a "diamond in the rough," though the sharper edges gradually eroded under the local social pressure. I made some further minor improvements and, supposing the primary cosmic rays were positrons and electrons (Caltech gospel at the time), averaged over angles of incidence to get the theoretical transition curve in the upper atmosphere for 11 GeV (then called BeV) primaries. The result looked quite similar to the San Antonio–Madras difference curve of Bowen, Millikan, and Neher. The theoretical curve died out by twelve radiation lengths; the experimental curve showed, by comparison, the presence at this depth of half a particle of secondary penetrating component per incident primary. The multiplication at the maximum of the shower curve was eleven theoretically and nine experimentally, a difference to be expected if energy is going into the production of the penetrating secondaries. Fortunately for me, the theoretical curve is not too sensitive to the fact that the primaries are protons, not positrons. That the primaries are protons was argued a year later, in 1939, by Johnson and Barry[20] and proved in 1941 by Schein, Jesse, and Wollan.[21]

There is a story connected to this paper which I owe to a historian of science (to whom I must apologize because I don't remember who it was). At the time I was working on the paper, Jim Bartlett, from the University of Illinois, was spending a sabbatical semester with us. He was engaged in two projects. The first was to learn Russian, which he attempted by comparing a Russian edition of Dirac's book with the English original. The second was to recalculate the screening constant that appears in the Bethe-Heitler formulae for bremsstrahlung and pair-production by high-energy electrons and gammas. The standard formulae had a screening factor in a logarithm of 183. Jim gave us a seminar on the subject and ended with a factor, which he said was preliminary, of 191. So in my shower theory I used 191 as the latest value, ascribing it to Bartlett. Lev Landau, in Russia, picked up the 191 from my paper and used it in his subsequent work. According to my historian friend the joke is this: In the years since, there have been two or three American projects to recalculate the screening constant, which all came up with Bethe's 183, and two or three Russian efforts, which checked Landau's 191.

Another paper appeared shortly, this one with Oppie, entitled "On the Stability of Stellar Neutron Cores" (19). In Pasadena Oppie was exposed to the staff of the Mount Wilson Observatory and to his friend Richard Tolman, and he had been interested in astrophysics for some time. Elis Strömgren's work on stellar models inspired us to try to work out the carbon cycle as the source of solar energy but we failed, misled by some erroneous experimental information, and Bethe scooped us. So this was our second try. Landau had suggested that main sequence stars might have a neutron core.[22] We said that Landau had greatly underestimated the minimum size the core could have, and correcting the estimate gave a minimum mass so large that the stellar model would be radically changed. We asked if inclusion of n-n forces could reduce the mass sufficiently to change the conclusion and found that it wouldn't with reasonable forces. The following year (1939), Oppie wrote "On Massive Neutron Cores" with George Volkoff[23] and, with Hartland Snyder, "On Continued Gravitational Contraction,"[24] the first description of a black hole. I was in on the initial hand-waving discussions of black holes but left Berkeley before the work had progressed very far.

At Berkeley we were diverted from the problems of electrodynamics by our interest in nuclear and particle physics. Meanwhile, at Stanford, Felix Bloch was working on electrodynamics, as well as doing experimental work in nuclear physics. In 1937 Bloch and Nordsieck explained the infrared divergence.[25] In 1939 Sid Dancoff got his degree with Oppie and went to Stanford as a postdoc. Felix put him to work calculating the ultraviolet vertex corrections. Sid showed me a manuscript of his paper before he published it and I found a mistake. He didn't correct the mistake, but mentioned it in a footnote and said it didn't really affect the conclusion.[26] Later Felix said to Oppie, "If Sid had gotten the right answer, he might not have known what it meant, but we sure would have!" (meaning they would have seen the solution to the renormalization problem then and there).

My paper on neutron cores was the last I would write from Berkeley (Oppie mailed it at the beginning of September 1938). In the spring of 1938 I was offered an assistant professorship at the University of Illinois in Urbana. At first I turned it down and Oppie nominated Leonard Schiff for the job. But then Rabi appeared and persuaded me to take the job. He told me jobs were very scarce. He told me it was doubly hard for a Jewish boy; he said that he himself was the first Jew appointed to the Columbia University faculty, and he said, anyway, to cut the umbilical cord. As a matter of fact, Oppie had tried for a couple of years to get me an assistant professorship at Berkeley, without any success. Many years later, I learned that Birge, the department chairman, had said in a letter that one Jew in the department was enough. Charlotte sided with Rabi, and I reluctantly gave in.

At the beginning of September 1938, we left Berkeley for Urbana. We packed all our belongings in our car, which was now a Model A roadster which we had bought from Stuart Harrison, and took a long way around. First, we drove up to Vancouver, to attend a meeting of theoretical physicists. A lot of our friends from Caltech and Berkeley were there, and we had a great time, both at the meetings and exploring Vancouver. Charlotte bought a beaver coat in preparation for living in a colder climate. The entertainment during that conference included a boat ride among the islands for all participants. When the time came to depart, a thick fog had rolled in and the visibility was limited to about fifty feet. The boat took off anyway, and we were fascinated by the way it was navigated. The pilot would blow a whistle, and by listening carefully to the way the sound echoed off the steep bluffs of the islands he could tell where he was. We were all sitting around on deck, and at one point someone asked what the consequences for physics would be if this boatload of theorists sank. Oppie instantly replied, "It wouldn't do any permanent good."

From Vancouver we drove east across Canada—slowly, because the roads were pitted with potholes. We went to Lake Louise, stayed for a couple of days there, scrambled around on the glaciers, then continued east until finally we got back home to Philadelphia. We hadn't seen our families in four years—not since we had left for

California in 1934. My father and stepmother were now living in a pleasant apartment in the center of the city. He had recently been appointed assistant city solicitor of the City of Philadelphia. My father told me that in the seventeenth century the city solicitor got no salary but was paid a fee for each service rendered. This system was never changed, and the city's population having increased a thousandfold, so had the fees. Despite customary kickbacks to the currently incumbent political party, the city solicitor's job was still a very lucrative one. From other sources I heard that my father had been the right-hand man of the winning mayoral candidate in the last election.

At the Leof house one afternoon, we met a very attractive girl, Kitty Puening, a biology student at the University of Pennsylvania. I suppose she had gotten to the Leof house through ties to the Medical Aid Committee; the previous year her husband had been killed fighting in Spain. It turned out that we'd overlapped for a year at the University of Wisconsin in 1933–34, and though we hadn't met, we had a number of mutual friends. But there was a more remarkable coincidence—she was engaged to marry Stuart Harrison, whose ex-car we had just parked in a garage around the corner. She had known him in England some years before and he had asked her to marry him then, which she had refused. Just a few months ago she had met him again in London, and he had renewed his proposal. This time she had accepted. They were to be married in Philadelphia in November.

After a week in Philadelphia, we turned our car west and headed, with some trepidation, for a new career in Urbana.

THREE

Urbana, 1938–1942

Urbana's Physics Department was lively. Wheeler Loomis, its chairman, had gotten money to build up his department. Jerry Kruger and Ken Green were building a cyclotron. Ed Jordan, a spectroscopist, came in 1937, as did John Manley, who began building a Cockcroft-Walton accelerator for nuclear physics experiments; he was joined in 1938 by my old Wisconsin colleague, Lee Haworth. Other new arrivals in 1938, besides myself, were Don Kerst from Wisconsin, Ernie Lyman and Reg Richardson from the Berkeley Radiation Lab and Maurice Goldhaber, a nuclear physicist from the Cavendish. Maurice, a refugee, was not Maurice in those days, but Moritz. In the summer of 1939, Maurice traveled to London to marry Gertrude Scharff, herself a very able nuclear physicist, and upon their return in September "Trude" came to the department free because of the university's strict nepotism rule. Loomis, who had four daughters, used to insist that if one of them married an Urbana staff member, it would be the son-in-law who had to leave. There was a remarkable statistic: over twenty girls born to Urbana physics professors in a row, until the string was broken by the Goldhabers on July 4, 1940. But Loomis claimed that didn't count because Freddie had been conceived before the Goldhabers arrived in Urbana. Maurice tried to save the rule another way, saying that one physicist makes a boy, two physicists make a girl. Later on I persuaded Loomis to add two more theorists as assistant professors, both Oppenheimer students and friends and colleagues from Berkeley—Sid Dancoff in 1940 and Philip Morrison in 1941.

Wheeler really had a buyer's market: jobs were still quite scarce in those post-Depression days, as Rabi had impressed on me. Once, sometime in the 1980s, I was in Rabi's office when he was looking for a document on the large table in the center of his room which was piled a couple of feet deep with assorted paper. A search was like an archaeological dig. After a while Rab said, "What's this?" and came up with a letter in his hand. It was from Oppie, undated, but the contents put it at 1940. It said, roughly, "Dear Rab: Can you help? I can't find a job for Julie Schwinger. The only opening I've heard of is at UCLA and the trouble is it came through Birge. Birge doesn't like Julie and will probably recommend Leonard Schiff or Hartland Snyder." Julian did get a job, at Purdue, where Frank Carlson looked out for him by making sure Julie, a nightowl, was awake to teach his afternoon classes.

Charlotte and I, however, were not enthusiastic about living in a small midwestern town. Urbana was tempered by the presence of a university. Still, the prejudice and intolerance of a small town were there. If you ordered something from the liquor store, the delivery truck was labeled "Candy." One evening in 1940 we were in a restaurant with the Krugers. At the next table a group of local businessmen were drinking and noisily discussing the upcoming presidential election. One of them turned to our table and demanded to know whom we were voting for. When I said "Roosevelt" he turned back to his friends and declaimed, "I told you every Jew in the country would vote for Roosevelt!"

Urbana summers were very hot—which didn't bother us because we were never there—and winters very cold. One Thursday the temperature was eighty degrees in Urbana when we left to drive to Chicago for the American Physical Society's Thanksgiving meeting. When we drove back on Saturday afternoon, bringing Victor Weisskopf with us, it was minus twenty degrees. After we entered the house, Viki sat down at the grand piano and began playing with gloves on, I went down to the basement to get a fire started in the coal furnace, and Charlotte went to the bathroom as one does after a long auto ride. She was surprised to see a yellow flood spreading across the tiles of the floor, freezing as it went. The water in the toilet bowl had frozen, cracking the bowl. Luckily, no water pipes were yet frozen.

Shortly after we arrived in Urbana, I received a call from my brother William, telling me of my father's death. He had been suffering from cancer and subject to spells of dizziness or faintness. In any event, he jumped or (less likely) fell from the balcony outside his office window.

At Urbana I directed the research of a few graduate students and taught two graduate-level courses a semester—six hours a week of lecturing (a heavy load compared to the postwar standard of one course and three hours). I taught a quantum mechanics course modeled on Oppie's. I developed other courses as well, including one on electricity and magnetism and another on nuclear physics. I had a mixed record as a teacher; I was good at lecturing, but bad at marking papers and getting them back on time. My biggest weakness was an inability to remember students—I could never seem to match names and faces. In my whole post-teaching-assistant career, no one ever trusted me with an undergraduate course.

Aside from my new colleagues at Urbana, I discovered that a number of young theorists were scattered among the universities of the Midwest, and we organized a permanent floating theoretical seminar which met once a month in Urbana, Bloomington, Lafayette, St. Louis, South Bend, or Evanston. We would drive from our home bases to the meeting place, often a two-day expedition. On one of these trips I learned about the flatness of the Illinois prairie. It had rained for days and the countryside was flooded. For mile after mile we rode along with the water level six inches below the surface of the road. I went on a trip to Purdue as a passenger in Maurice Goldhaber's car. He had learned to drive only since arriving in Urbana, and his technique was to floor the accelerator and let her rip. On the way over, we descended an Indiana hill. We saw that it was crossed at the bottom by a railroad track occupied at the moment by a moving freight train. Maurice aimed to pass just behind the last car. Almost at the last moment we heard the whistle of another train, approaching the crossing in the opposite direction on a second track and hidden from us by the first train. On the way home, late that night, Maurice passed a car just before we reached the crest of a hill. When I remarked that this was dangerous he answered, "If a car were coming with headlights on I'd see the scattered light." These theoretical seminars continued until the United States entered the war.

One of the first things I did after my arrival at the University of Illinois was to write a letter to the dean of the Graduate School sup-

Fig. 3.1 Four future presidents of the American Physical Society (including two future Nobel laureates) play hooky from the June 1938 APS meeting to visit the San Diego Zoo. *From left to right*: Robert Serber, Willy Fowler, Robert Oppenheimer, and Luis Alvarez.

porting Don Kerst's application for research funds to build an electron accelerator. The sum he was asking for was $400. Don called the machine an "induction accelerator." Later on it picked up the name "betatron," a name Don never liked, but one that stuck. Don and I collaborated in calculating the orbits of electrons in the time-varying magnetic field. In addition to studying the focusing conditions, we gave careful thought to how electrons could be injected into the machine. This, plus Don's experimental talents, would lead to success where earlier attempts had failed.

While I was in Berkeley in September 1940, Ernest Lawrence got a message from Don that his prototype betatron (2.3 MeV energy, 7.5

cm radius) was working, and that same evening, at Ernest's Journal Club, I gave a report on what it was. In 1941 Don and I published a paper on the theory of the betatron, "Electronic Orbits in the Induction Accelerator" (22), a paper that laid a basis for the design of future circular accelerators.

The paper originally had a final section, supplied by Don, on what is now called "synchrotron radiation," which gave the energy lost per turn to radiation and said this limited the energy a betatron could reach. Although the theoretical paper was the logical place for this section, I complained to Don that it was his work, not a joint effort, and asked him to move it to his experimental paper, which preceded ours in the *Physical Review*. The upshot was that somehow it got left out of both papers.

At the time they were published Don was in Schenectady at the General Electric Research Laboratory building a 30-MeV betatron. Some GE people argued that there wouldn't be any synchrotron radiation because the circulating electrons amounted to a ring of direct current. I looked into it and showed that Don's energy-loss formula was correct, the radiation was due to density fluctuations in the ring. This calculation was a partial and incomplete version of calculations earlier by Oliver Heaviside and later by Schwinger. Don suggested that an oscillating electric field could make up the radiation loss, and I was just beginning to look at this suggestion—which might have led us to phase stability—when Oppie dragged me off (in 1942) to work on atomic bombs.

Oppie would write me every Sunday. From one of those Sunday letters, which I received in January of 1939, I learned of the discovery of fission. In that first letter Oppie mentioned the possibility of nuclear power and of an explosive. My immediate reaction, and I'm sure that of most other nuclear theorists, was that I should have thought of fission myself. I'd heard Glenn Seaborg talk about the troubles with the chemistry of the supposed transuranics. I went to the library and found an article in the *Handbuch der Physik* on the oscillations of a liquid drop. I had only to add an electric charge to the drop. When I tried the crude criterion that fission would occur when the amplitude of oscillation became equal to the radius of the drop, I found I had a reasonable model for nuclear fission. I talked about fission that evening at the Journal Club. Maurice Goldhaber heard the news the same day from friends in the East.

Fig. 3.2   Charlotte and I, with Ed and Elsie McMillan (back couple).

Summers we spent in the West, dividing our time between Pasadena, Oppie's ranch, and Berkeley. In Pasadena in the summer of 1939 we met Kitty again, recently graduated from Penn, married to Stuart Harrison and now part of the Caltech community. Then to the ranch. During our stay, Oppie and I went to one of the few conferences that we attended in those days, a cosmic ray conference in Chicago. We drove there and back, and on our return were sternly cross-examined by Charlotte as to whether we had eaten any vegetables during the entire trip.

After the ranch we returned to Berkeley where the term began in mid-August, almost a month and a half before Urbana. During this interval in 1939, Oppie and Hartland Snyder and I wrote a paper entitled "The Production of Soft Secondaries by Mesotrons" (21). Remarkably, this title didn't begin with "On" or "Note on," as did many Berkeley papers. In my December 1937 paper on the origin of nuclear forces, I called Yukawa's particle a "dynaton" and later, I remember we used "Yukon." Now it's become "mesotron," on the way to "meson." As far as length went, this reversed the evolution of "deuteron," which in Berkeley in 1934 we were calling the "deuton." When the Cavendish started using "deuteron" the joke went around that Rutherford wanted his initials included. In our 1937 "Note on the Nature of Cosmic Ray Particles" (12), Oppie and I had stated that knock-on electrons accompanying the penetrating component accounted for the showers observed below sea level. In the new paper we gave the details of the calculation for a spin ´ particle. The knock-ons accounted for the soft component up to 20 GeV. Above 20 GeV the "bursts" observed under lead absorbers at sea level could be accounted for as due to the bremsstrahlung of the penetrating particles.

However, we thought the penetrating particles were Yukawa mesons, and that nuclear forces required them to be spin 1. But when we calculated knock-ons and bremsstrahlung for spin 1, we found we overdid it. In fact the bremsstrahlung increased so much with energy that the high-energy penetrating particles would not be penetrating. Instead of drawing the correct conclusion, that the penetrating component had spin $\frac{1}{2}$, we suggested that the calculated cross sections for spin 1 were so large that perturbation theory was unreliable. Shortly thereafter, more careful treatments by Oppie's students Christy and Kusaka[1] and Corben and Schwinger[2] showed that indeed the spin could not be greater than $\frac{1}{2}$.

Later that fall, back in Urbana, I examined another aspect of Yukawa's theory, and showed that if the Yukawa particle had spin 1 one had to give up Yukawa's original suggestion that beta decay went through virtual mesons (20).

When we reached Berkeley in the spring of 1940, Oppie had returned from Pasadena and was about to leave for the ranch. Charlotte and I decided to go the long way round and visit Pasadena on our way to New Mexico. Oppie told us that he'd invited the Harrisons but that Stuart couldn't get away. Then he said, "Kitty might come alone. You could bring her with you. I'll leave it up to you. But if you do it might have serious consequences."

We brought Kitty with us. A day or two after we arrived, Oppie and Kitty rode down for an overnight visit to Los Pinos, a dude ranch in Cowles owned by Oppie's old friend, Catherine Page, with whom he had stayed when he first went to New Mexico for his health. The next day, after they returned, Catherine, looking very aristocratic on her bay horse, came trotting up to the ranch house and presented Kitty with her nightgown, which had been found under Oppie's pillow. The rest of us made no comment. However, that afternoon Charlotte and Frank's wife Jackie went for a ride and when they returned Jackie, who was on the lead horse, had a stiff neck from conversing over her shoulder.

In October 1940, back in Urbana, we received a surprise phone call from Oppie, who was in Chicago on his way West. He said he had something to tell us and asked us to come meet him in his hotel room. When we arrived he sat us down. "I have some news," he said, "I'm going to marry—" and at that moment he dropped his voice, a trick to get attention he learned from Bohr. I just gaped at him, trying to figure out whether he'd said "Kitty" or "Jean." Charlotte had to kick me to remind me to make the appropriate salutary noises.

Kitty, it turned out, was then in Reno establishing residency for a divorce from Stuart Harrison; Oppie was on his way to meet her there. A few days later, on November 1, 1940, Kitty was divorced from Stuart and married to Oppie.

Meanwhile, I began working with Dancoff on nuclear forces, trying to make use of an idea of Heisenberg's. In 1939 Heisenberg, trying to resolve the supposed mystery of the small scattering of the mesons of the penetrating component, argued that the fault lay in treating the nucleon-meson coupling as weak: taking the example of a neutral pseudoscalar theory, he said the coupling would introduce an inertial term in the equation of motion of the spin, which would lead to a reduced scattering.[3] In 1940 and 1941 Gregor Wentzel introduced strong coupling theory and found charge isobars for the charged scalar theory.[4] In 1941 Oppenheimer and Schwinger did the strong coupling calculation for symmetrical scalar and neutral pseudoscalar theories and in the latter case indeed got a small scattering cross section.[5] At the same time, at Illinois, Sid Dancoff and I were learning about the same two theories. It had occurred to me that Heisenberg's effect would help nuclear force theory, damping the $1/r^3$ radial dependence of the dipole-dipole type tensor force. But we found that strong coupling only made things worse; the radial dependence was unchanged and as soon as the nuclear interaction became larger than the isobar or isotope separations the forces became an ordinary central force, non-exchange and spin independent.

We first reported this early in 1942 at a meeting of the American Physical Society, and the abstract was published under the title "Nuclear Forces in the Strong Coupling Theory" (23). Sid was supposed to write up the final version of the paper, but before he got around to it he got an appointment to work with Pauli at Purdue. Pauli found out what he was doing, became interested, and began to work with Sid on the symmetrical pseudoscalar theory, a more complicated case. I complained to Oppie that Pauli was working Sid so hard that he hadn't had time to complete our manuscript, and Sid's paper with Pauli, properly a sequel to ours, would appear first. As Sid was picking up his mail in the hallway of the Purdue physics building shortly thereafter, he pulled out a letter from Oppie. Pauli happened to be right beside him, and rudely read it over Sid's shoulder, seeing Oppie's comment that Pauli had a sufficient reputation that he didn't have to horn in on the credit of younger colleagues. Pauli was furious but not contrite: the Pauli-Dancoff paper did appear before the Serber-Dancoff one (24). Later it was found that the pi meson is indeed pseudoscalar and the Pauli-Dancoff prediction of a $T = \frac{3}{2}$, $J = \frac{3}{2}$ resonance was borne out by Fermi's pi-nucleon scattering experiment.

In June of 1941, Charlotte and I drove to the East Coast to see her family. We had car trouble, and when we made it back to Urbana at the beginning of July we bought a new car, a black Oldsmobile convertible—we discovered belatedly that black is harder to keep looking clean than white—and proceeded to the ranch, arriving unusually late, in mid-July. It was a bad-luck summer. One day in the corral, Charlotte's horse kicked Oppie on the knee. It didn't break anything but left a painful bruise. And a couple of days later Kitty, driving her Cadillac convertible between the ranch and Santa Fe, had to slam on her brakes to avoid hitting a car that stopped suddenly in front of her and was thrown forward, hurting a leg. The only cheerful occurrence was Oppie's running into Hans Bethe walking in the neighborhood and bringing him up for a visit. Kitty and Oppie couldn't ride, so the ranch was boring. Also they had bought a house in Kensington, just north of Berkeley, and were anxious to move in. They left at the beginning of August. Jackie was not displeased; she disliked Kitty. When together the two tended to play the roles of working girl and aristocrat, and looked down at each other. We stayed and took a six-day pack trip with Frank and Jackie before returning to Berkeley in the middle of August. We helped the Oppenheimers move into their new home at Eagle Hill Road. The one-story Spanish-style house had the feeling that many Berkeley hill houses achieved of being quite isolated from its neighbors. It was perched on top of a little knoll, Eagle Hill, with a garden behind and a garden in front. Before that was a garage with a room and bath above, and the driveway down to the street below. From the front garden one had a wide view of San Francisco Bay.

We returned to Urbana in mid-September of 1941. I remember hearing of the Pearl Harbor attack on the car radio while driving back to Urbana from a seminar at Purdue.

PART II

War

. . . .

FOUR

Berkeley and Los Alamos, 1942–1945

About Christmastime in 1941, just a few weeks after Pearl Harbor, I received a phone call from Oppie. He said he was in Chicago and wanted to come down and talk to me about something. He came, and we took a walk out to the cornfields beyond the edge of town. There, alone in that rural setting, he told me that he was going to be appointed to head the weapons end of the atomic bomb project, to replace Gregory Breit in that position. He wanted me to come to Berkeley and be his assistant in the project.

There was some difficulty in leaving Urbana immediately. Many of the staff had already left for war work, including our chairman, Wheeler Loomis, who was now associate director of the Radiation Laboratory at MIT. There was no one to take over my teaching for the coming semester. Things were so tight that Sid Dancoff was dragooned into teaching the large elementary physics course for engineers. He'd never done anything like it before. At his first lecture he mounted the rostrum, turned and looked out over a sea of four hundred eager young faces, and fainted dead away. So it was agreed that I would come as soon as the semester finished at the end of April. I didn't know it at the time, but Oppie's appearance and prospective appointment were the result of an upheaval and a change—a major change—in American policy toward the uranium project.

Everyone has heard about the letter that Einstein sent to Roosevelt in October 1939 that supposedly set in motion the events that led to the uranium project and eventually the Manhattan Project. Rabi said that letter actually held things up for a year. He said that Leo Szilard didn't know the ropes, didn't know how to make things happen. For instance, while he and Rabi were in the same department, and the same laboratory, Szilard never spoke to Rabi about it. Be that as it may, the result of that letter was that the project got pushed into a bureaucratic backwater. It was turned over to Lyman Briggs, the director of the National Bureau of Standards, who appointed Gregory Breit, a theorist from the University of Wisconsin, to be in charge of the bomb part of the project. For the next two years they did practically nothing. Breit started a few projects to measure high-energy neutron cross sections at universities doing nuclear physics, to the tune of about $40,000 a year. Briggs gave $6,000 to Fermi and Szilard to buy graphite and uranium oxide. He gave nothing at all to John Dunning and Eugene T. Booth for their gaseous diffusion experiments at Columbia. The project was limping along and, in fact, Vannevar Bush, the chairman of the Office of Scientific Research and Development, was thinking of closing it down completely, until the British came along and saved the day.

Rudolf Peierls and Otto Frisch, in Birmingham, wrote a memorandum to the British government in March 1940, saying that a fast neutron uranium 235 fission bomb was practical and that it was possible to separate the two isotopes of uranium, they thought by thermal diffusion. Their optimism was enhanced by a miscalculation of the critical mass, the minimum mass required to sustain a chain reaction; they underestimated it by a factor of twenty-five. They thought it was 600 grams of uranium 235. Politically, this was probably a fortunate mistake. It may have helped get the British government interested. A committee, the so-called MAUD Committee, was appointed, which, in the summer of 1941, reported that a bomb was indeed feasible. They raised the critical mass to somewhere in the neighborhood of three kilograms, and they said that it could be built in two and a half years, at a cost of about £5 million. They said, "In spite of this very large expenditure . . . every effort should be made to produce bombs of this kind." The British estimate of cost was even worse than their estimate of critical mass; here they were low by a factor of one hundred.

The report of the MAUD Committee was officially transmitted to the American government in October 1941. It finally persuaded Bush and James B. Conant, the chairman of the National Defense Research Council, to proceed, with high priority, on the uranium and plutonium project. They met in Washington the day before Pearl Harbor, December 6, 1941, with Arthur Compton. It was decided to take the project out of the inept hands of Briggs and Breit and turn it over to Compton, who would be responsible for the reactor work and bomb design; to Ernest Lawrence, with his electromagnetic separation project in Berkeley; and to Harold Urey, representing the gaseous diffusion project at Columbia. Fermi's Columbia operation was put under Compton's authority.

Lawrence had brought Oppenheimer to a meeting called by Compton in Schenectady in October. Compton, who had failed to get any information or answers out of Breit, turned to Oppenheimer and was impressed by the help he received. So, in the course of the reorganization, he decided to replace Breit with Oppenheimer, although he didn't do it immediately but eased him out gradually over the course of the following spring. Thus the call from Oppie for my help in Berkeley.

About the end of April Charlotte and I packed up our car and started for Berkeley. I don't remember how we got the gas coupons for the trip; it must have been a special dispensation of some kind. Remember, everything—gas, meat, cigarettes, liquor, all the necessities of life—was strictly rationed in those days. We got to Berkeley and discovered that the quiet university town was quite altered. The Richmond shipyards were running full blast, and housing was almost impossible to get, it was so crowded. Temporarily, we moved into a room over the garage at Oppie and Kitty's house at 1 Eagle Hill, just north of Berkeley, and we never got out of it the whole year we were in Berkeley. There was some paranoia in the area; they enforced a blackout at night, and cars had their headlights painted black with just a little slit to let a bit of light out. On the other hand, from Eagle Hill, the whole sky was lit up by lights from the Richmond shipyard, which if anything was the main target in the area.

Charlotte got a job at the shipyard as a statistician, which just meant that she told them that she knew how to add and subtract. Her boss was very pleased to have someone who knew how to add and subtract, and in a short time she found herself in effective command of a payroll of about six million dollars a week while her boss goofed off.

The day after arriving in Berkeley, I went down to Oppie's office in Le Conte Hall where he had accumulated a number of British documents concerning bomb design. I remember there was a paper on critical mass and something on assembling the pieces of a bomb. Perhaps there was something on efficiency, I don't remember. The papers were rudimentary but were really quite helpful in getting us started. For example, in the critical mass calculation, what you have to know is the chance that a neutron made somewhere inside the bomb's sphere will escape through the surface. Moreover, if the bomb's sphere is surrounded by some reflective material (a tamper we called it), what percentage of the neutrons that escape are reflected back? Now, the elementary theory that was used in the British papers was based on an approximation, which was that the neutron made many collisions before it escaped through the surface. This condition was certainly not met in the problem we were considering, and there was a large question of what effect this would have on the estimated critical mass.

Oppie had a team of young postdocs and graduate students who were working on calculations for Lawrence's electromagnetic separator, and I had their use, as long as I didn't take up so much time that Ernest's requirements suffered. I suggested to two of the members of that team, Eldred Nelson and Stan Frankel, a way of improving this elementary calculation of the critical mass. They took up the problem and went off on their own and did considerably better than what I had suggested. They wrote down an exact equation for a diffusion problem, found properties of its solution, found in the literature a case where it had been solved exactly, and put themselves in a position to make quite accurate calculations, provided, of course, one knew the physical constants, such as the value of the cross sections and the number of neutrons per fission.

For uranium, we had reasonable values for the constants. For plutonium, not much. Glenn Seaborg and Emilio Segrè had used the microscopic quantity of plutonium made on the Berkeley cyclotron and determined that the fission cross section for plutonium was about twice as large as for uranium. For the other constants, we just guessed they were the same as for uranium. The improvement in the calculations over the simple theory reduced the critical mass for a bare uranium sphere by more than a factor of three and for a tamped sphere by almost a factor two.

We finally came up with a calculated critical mass of 15 kilograms for uranium and 4 kilograms for plutonium, which turned out to be about the right answers. Our 15 kilograms for uranium is to be compared with the 600 grams of Peierls and Frisch, while the Germans seemed to think it was about a ton. A remarkable thing is that there was no American estimate prior to the MAUD report. If anybody had given the matter any thought in this country, there was no record of it. If Breit knew anything he kept it entirely to himself, which is the second-best security. The best security is not to know anything in the first place.

While Nelson and Frankel were calculating the critical mass, I was looking at the efficiency problem, which involved not only neutron diffusion but the hydrodynamics of the explosion. When the bomb begins to explode it blows itself up, as well as its surroundings. The expansion stops the chain reaction before all the fissionable material is used up. The efficiency is the fraction which fissions, and determines the energy release.

Oppie, who had inherited the cross-section measurement projects set up by Breit at the various universities, was away a good part of the time getting them reasonably organized and correcting the morale problems resulting from Breit's excessive secrecy. He also was getting information on subjects with which we were quite unfamiliar, like damage caused by high explosives and types of guns available from the army and navy.

Guns came into it because to set off the explosion two subcritical pieces had to be brought together to make a supercritical one, and it had to be done fast enough so that some stray neutron didn't set off a chain reaction while the pieces were still not in place, which would lead to a low efficiency and a much smaller yield of energy. That was a real problem, particularly for plutonium since plutonium is quite radioactive and emits a lot of alpha particles which can make neutrons by nuclear reactions in light elements which might be present as impurities. Even with the fastest guns available, it turned out that you needed purities on the order of parts per million of a number of light elements, a very tough job for the chemists and metallurgists.

I didn't notice it at the time, but the security services were keeping an eye on the Berkeley projects. One day Martin Kamen, one of Ernest's boys and the discoverer of carbon 14 (which revolutionized archaeological dating), was indiscreet enough to have lunch in San Francisco with an attachè from the Russian consulate. A row devel-

Fig. 4.1 Original badge photograph of Edward Teller (1908—) at the Los Alamos Scientific Laboratory (it is unclear what "U 10" stands for). Teller was born in Budapest, Hungary, in 1908 and received a Ph.D. from the University of Leipzig in 1930. He was a professor of physics at George Washington University from 1935 to 1941, and at Columbia University in 1941–42. He worked for the Manhattan Engineering District at the University of Chicago in 1942–43, and at Los Alamos from 1943 to 1946. He returned to Chicago in 1946 to take a position with the Institute for Nuclear Studies and remained there until 1975. That year, he became a senior research fellow at the Hoover Institute at Stanford University. (Photo courtesy of Los Alamos National Laboratory)

oped in the restaurant, agents of Army Intelligence, Navy Intelligence, and the FBI arguing about priority: who would get to occupy the next booth.

In July 1942, after we'd been working for two months, Oppie called a meeting of theorists in Berkeley to discuss the feasibility of a bomb. There were Hans Bethe, Edward Teller, John Van Vleck, Richard Tolman, Felix Bloch, Konopinski, Frankel and Nelson, Oppie, and myself. I led off, telling what we had done, and Frankel and Nelson described their calculations. Everybody agreed that it looked under good control from a theorist's point of view, the main problem being the assembly problem for plutonium. For instance, it would need a high-velocity gun, and one of these weighed thirteen tons, a little much to carry in a plane, though Oppie remarked it might be lightened some because it only had to be fired once.

We talked about damage: from the blast, from neutrons, and from radioactivity. We talked about nonspherical shapes and various fancy so-called autocatalytic schemes which Teller brought up, in which neutron absorbers are compressed by the developing explosion, and which turned out to have too high critical masses and too low efficiencies to be interesting.

At that point something remarkable happened. Teller brought up the idea of the Super, a fusion weapon, not a fission weapon, which was to be a detonation wave in liquid deuterium set off by being heated by the explosion of an atomic bomb. Well, everybody forgot about the A-bomb, as if it were old hat, something settled, no problem, and turned with enthusiasm to something new. At first Edward claimed it was a cinch until somebody pointed out cooling by radiation. Then Edward came back and said the cooling would be too slow to do any harm. And the next morning Hans came back with a nice fast mechanism for cooling. So it looked as if the Super were dead until K'ski (Konopinski) breathed a little life back by suggesting the addition of tritium. So that's the way it went, day to day, back and forth, everybody having a great time with the idea. At one point Edward asked if the fission bomb could ignite the earth's atmosphere. In view of the difficulties encountered in considering the Super this seemed extremely unlikely, but in view of the importance of the consequences, Hans took a look at it and put numbers to the improbability. But meanwhile Oppie told Compton and Compton told Washington and thereafter, every once in a while, someone would ask, "Are you sure?" Later on K'ski did some more work on the question.

While this was going on, Richard Tolman, who was not a nuclear physicist, came to me and said we really should think more about implosion, that is, assembling the pieces by use of high explosives. We talked about it, and Richard and I wrote a memorandum on the subject. Later on Richard pursued it, which he was in a good position to do as vice chairman of the National Defense Research Council. Science historian Lillian Hoddeson told me that in the records there are two more memos about implosion written by Tolman, and minutes of a meeting in which Compton and Bush urged Oppenheimer to pursue the implosion idea. He replied, "Serber is looking into it."

The conference didn't exactly end, it sort of fizzled out. After a week people began to leave, some stayed on a couple of weeks longer, and then, during the fall, Oppie and I went up to Chicago a couple of times to talk to Bethe and Teller and Konopinski about what we were doing and what they were doing—various parts of the problem like equations of state and opacities.

On one of these trips we were present when Gen. Leslie Groves first met the scientists of the project. Before the meeting Compton was running around looking for Szilard; he wanted to keep him away from Groves. Szilard's mannerisms gave somewhat the impression of the "mad scientist." He would press his latest idea upon anyone who would listen, and I was told his current one was to use radioactive material produced in nuclear reactors to make flashlights for our soldiers. We all had our idiosyncrasies, but Szilard's were more obvious than most. Finally we were all sitting around a big table when Groves came in and we were each introduced to him. He started to give a speech along the lines that "You're working for me now and you'd better toe the line" when, in about two minutes, he was called away to the phone and never came back.

About three days later, back in Berkeley, Oppie and I were working in his office when Groves came in, followed by Col. Kenneth D. Nichols. The first thing Groves did was take off his jacket, hand it to the colonel and say "Find a tailor or dry cleaner and get this pressed," and Nichols took it and walked out the door. I was quite impressed by the way a general could treat a colonel. But that was Groves's way; I think it was a matter of policy to be as nasty as possible to his subordinates. Priscilla Green Duffield, Oppie's executive secretary, had a story of a typical Groves visit to Los Alamos: he would stride into her office, heading straight for Oppie's door, and in passing glance at her and say, "Your face is dirty." But I must say he did a great job as head of the Manhattan Project. There is hardly a better indication of this than his unlikely choice of Oppenheimer to be the director of Los Alamos. On another Chicago trip we visited Fermi's reactor about a week before it got going.

One evening, driving back to 1 Eagle Hill from the office, I was surprised to see Oppie and Jean Tatlock pacing along the sidewalk below the house. Later, Kitty told me that still, when Jean was hit by a bad depression, she would appeal to Oppie for support.

During the fall and winter we continued our calculations. For example, in the critical mass calculation we had made a reasonable guess as to the mean velocity of the neutrons. We now improved the reliability of the calculation by considering several groups of neutrons of different velocities. I worked out two problems in connection with the hydrodynamics of the explosion. One was the effect on the efficiency of a bare sphere of the rarefaction wave at the sphere's surface. The other, needed for a tamped sphere, was the theory of the exponential shock wave, a shock wave with pressure increasing exponentially with time. In the end it required numerical integration of a differential equation, a job undertaken by Frankel and Nelson, who were much surprised when it turned out to be much more difficult than they expected. It needed higher accuracy than they had anticipated. All our calculations, you must realize—it's hard these days to think of it—were done with mechanical desk calculators, Marchands and Monroes. That problem got Frankel and Nelson interested in computing, and they pursued it later on in Los Alamos, where they finally became the group leaders for the implosion calculations by state-of-the-art methods, which in those days meant using IBM punch card machines. While Frankel and Nelson were struggling with the numerical integration, we received a classified paper by Dirac on the same subject. However, it contained an error: Dirac's solution behaved incorrectly at the singular point of the differential equation.

In the meantime, Groves had decided to start a bomb laboratory, and all during the fall and winter Oppie was very busy getting it organized. He and Ed McMillan found a site in the Los Alamos Ranch School for Boys, which was just across the Rio Grande Valley

Fig. 4.2 Charlotte's staff at Los Alamos. To her right is Frances Hawkins (Dave Hawkins's wife), who briefly worked as a bookbinder (for security reasons, books were bound on site).

from Oppie's New Mexico ranch. The original idea was that there would be about fifty scientists, and the auxiliary jobs that had to be done, like secretarial work, would all be done by the scientist's wives. Oppie picked Charlotte to be the librarian. He thought she would do it more efficiently than a professional librarian, appreciating when to cut the necessary corners; Oppie liked to say that Charlotte was unfettered by professional librarians' dogmas like finding the names that belong to initials. Still, I don't think either of them realized at the time how big the job would be. Charlotte would wind up in charge not only of the library but also of the document room, containing all the secret documents, and of the publication of all Los Alamos reports. In a short time she had about a dozen people working for her. She was the only woman group leader at Los Alamos. For a while there was talk of our all becoming army officers, which bothered Charlotte because a WAC had to be five feet tall and she could only make about four eleven and a half. But Rabi and other experienced people talked Oppie out of that one.

In the beginning of March we left Berkeley and drove out to Santa Fe. There we turned north up the Rio Grande Valley—if we'd been going to the ranch we'd have continued east for another twenty miles and turned up the Pecos Valley—for about twenty miles, then turned left on a two-lane dirt road that dipped through half a dozen arroyos before reaching the Rio Grande at a place called Otowi. There the river was crossed by a toy one-lane suspension bridge that looked as if it might be safe for two horses. It was hard to believe that all the construction trucks for Los Alamos had to cross that bridge and then climb 1,500 feet up a perilous switch-backed dirt road to the top of Los Alamos Mesa, from which there was a great view, across the Rio Grande Valley, of the Sangre de Christo Mountains.

As one would expect, the site itself was a mess. When we arrived, the original school buildings were usable, but the technical area buildings weren't finished and neither was the housing. Charlotte and I were the first to arrive after Oppie and Kitty, who took the headmaster's house, and we were put up in the Big House, the dormitory for boys at the Ranch School. After we arrived, people with families were put down in ranches the army rented in the Rio Grande Valley and the single guys were given rooms in the Big House. The accommodations were a little rough; there was only one big bathroom in the entire house, and two or three fellows were embarrassed by walking in on Charlotte while she was taking a shower. We ate at Fuller Lodge, another school building. The food was good, but sparse; some of the younger and larger scientists, like Bob Cornog, claimed that they were starving and after meals would scrounge around for any food left untouched.

In a little while we were able to move into our apartment in a bungalow, a duplex with two apartments next to each other, we in one side, Bob and Jane Wilson in the other. A little primitive, all we had for cooking was a wood stove. Of course there were a lot of complaints which General Groves pooh-poohed, and on one of his first trips to Los Alamos he came over to our place, to Jane Wilson's kitchen, and pretty soon the general was on his hands and knees blowing on the kindling. At that altitude (7,300 feet), it's a little difficult to start a fire. As a result of his experience we were issued electric hot plates. One still had to get used to cooking there because at

Fig. 4.3 Site of the Los Alamos laboratory, taken during Ranch School days; Ashley Pond, with Fuller Lodge in the background.

Fig. 4.4 Bob and Jane Wilson, and Charlotte on the right.

Fig. 4.5   Charlotte at Los Alamos.

7,300 feet water boils at 197 degrees rather than at 212, and it takes five minutes to boil a three-minute egg.

No books had come to the library yet, so Charlotte worked for a while in Oppie's office, helping Priscilla Green Duffield. There was only one telephone line into the place at the time, the old ranger line that served the boys school. One day, when there was a thunderstorm down in the Rio Grande Valley, the phone rang and Charlotte reached to pick it up and a spark jumped from the phone to a lamp cord about six inches away. After that, no one would answer the phone if there was a cloud anywhere in sight. The ranch school had left a herd of horses, and Bob Wilson, a real horseman from Wyoming, mounted all of the boys he had brought from Princeton on horses, galloping all around the horse field in a big cloud of dust, Charlotte and I following behind watching his boys falling off right and left.

Not only in construction was nothing ready. No security arrangements were in place. Oppie wrote passes for us on University of

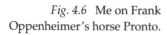

Fig. 4.6 Me on Frank Oppenheimer's horse Pronto.

California letterhead stationery which didn't well survive being carried in hip pockets. Spanish-American guards from the construction sites were dragooned to man the gate. One night Johnny Williams was driving Rose Bethe up from Santa Fe. At the gate the guard pretended to read his pass and, being a gentleman, ignored the fact that there was a woman in the car. The army shortly took over the horses and brought in MPs who were mostly ex–New York cops and put the New York cops on the horses—probably none of them had ever seen a horse before—and set them to patrolling the fences. So there was a lot of suffering for a while until they called that off after a couple of weeks.

There were, of course, a lot of rumors circulating in Santa Fe about what was going on up at Los Alamos Mesa, and Oppie and the lieutenant in charge of army security decided that the thing to do was to spread counterrumors. They asked Charlotte and John Manley to go down to town and spread the rumor that we were building an electric rocket. They weren't very enthusiastic about it, so Charlotte asked if I could go along and John said he would take Priscilla and the four of us went down to Santa Fe. We went to the cocktail lounge at the La Fonda, which was usually quite a lively busy place but on this particular evening was rather quiet. We got a table and drinks. We found it difficult to work electric rockets into the conversation. But nobody paid any attention that we could see, so after a while we left there and decided to go to a low-down bar. The bar was jumping, jammed and crowded and full of Spanish-Americans. We got a booth and drinks, and John and Priscilla started to dance and talk about electric rockets. Pretty soon a Spanish-American kid came over and asked Charlotte to dance. It turned out that he'd had a job as a construction worker on the site for a while and couldn't care less, and the only thing he wanted to talk about was his ambition to own a horse ranch. I saw things weren't going so well, so I went up to the bar and I grabbed a drunk by his lapels and shook him. I said, "Do you know what we're doing at Los Alamos? We're building an electric rocket!" The only trouble with that was he was so drunk I'm sure he didn't remember a thing the next morning. So our expedition was a flop, the FBI and Army Intelligence never reported picking up any rumors about electric rockets. Charlotte and John made one more effort: she went to a beauty parlor in town and he to a barber shop, but they were no more successful. The spy business isn't as easy as it appears in the movies.

Speaking of security, there were a couple of other amusing things about Los Alamos in the early days. All the outgoing mail was censored, but in the beginning it was not mailed locally but from Los Angeles, Chicago, or New York, and Charlotte had a fine time mystifying her parents by alternately putting her letters in the three different mailbags. One of my best souvenirs was my New Mexico driver's license. On the line that said "Name" it said "Not Required" and on the line that said "Address" it said "Special List B." Unfortunately I lost that when I was mugged once in Panama.

The Los Alamos security arrangements were another example of Groves's talent for running the project. If necessary, he was willing to be unorthodox. He made the laboratory responsible for its own security. In hiring this meant no delays waiting for clearance, and it avoided the security services prejudices against leftists and emigrés. And within the laboratory there was no "need to know" kind of rule; all information was free to all the scientific staff. We did some unprofessional things: can you guess what the penalty was for leaving a secret document on your desk at night? If you were caught, you were

It and you had to go around the next night and succeeding nights trying to catch somebody else. One day I was walking down the hall
and came on Dave Hawkins and Emilio Segrè having an argument.
Hawkins was in charge of running this program. He was saying,
"Emilio, you left a secret paper out last night and you have to go
around tonight." But Segrè was having none of it. He said, "That
paper, it was all wrong. It would only have confused the enemy."

A real security crisis occurred when we discovered that someone
was breaking into the lab refrigerator at night and stealing food.
Robert Wilson set up a flash camera that would go off when someone
opened the refrigerator door. The thief, of course, realized that his
picture had been taken, easily located the camera, and removed the
film. But Bob, of course, had already anticipated this and had set up
a second, well-hidden camera, set to go off at the same time as the
first. The culprit turned out to be a night watchman.

All during March staff was arriving, apparatus was shipped in
and set up, and at the beginning of April I gave a series of indoctri-
nation lectures to tell the staff what the project was about and what
we knew about it. The lectures were given in the library, which was
empty, for no books had arrived yet. We set up a little blackboard
with folding chairs in front of it and about thirty or forty people
came. Security was terrible, we could hear carpenters banging down
the hall and at one point a leg appeared through the beaverboard
ceiling, presumably belonging to an electrician working up above. A
couple of minutes into the first lecture Oppie sent John Manley up to
tell me not to use the word "bomb" but to use something neutral like
"gadget." I did, the word stuck, and after that the bomb was always
called the gadget at Los Alamos. I gave five lectures, an hour or so
each, over the course of two weeks. Ed Condon, who was the associate director of the lab, wrote up the notes of the lectures and
Charlotte published them as *Los Alamos Report Number 1: The Los
Alamos Primer*. During the last lecture, after I had mentioned implosion as a possible means of assembly, Seth Neddermeyer got up and
expressed the opinion that this was the way to go. He followed up by
starting an experimental program to study implosion.

One day while the lectures were going on Oppie asked me into his
office and told me he was organizing the lab, military-style, into divi-
sions (designated by letters), each composed of a number of groups
(designated by numbers). He said that Bethe would head the

Theoretical Division (T), while I would be a group leader (T-2). The notion of an organization was a shock; I had gotten used to working independently. But it was logical, and the fact that it was Bethe made it easier to accept. A couple of days later, Hans came into my office to discuss what my group, called "Diffusion Theory, IBM calculations, and experiments," would be responsible for. He had us down for diffusion theory calculations and for advanced calculations methods, which meant the use of IBM punch card machines, which we had gotten into through Frankel and Nelson's work on the exponential shock wave. I asked about implosion, which Oppie had told me to look at while we were still in Berkeley, but Hans said my group couldn't do everything and his own group would take care of implosion.

The notion that the lab would be only fifty people didn't last very long. All kinds of people were arriving: doctors for the hospital, nurses, schoolteachers, electricians, carpenters and machinists, who, by the way, were paid more than the scientists. The wives took jobs. Jane Wilson was a schoolteacher. Vera Williams, Johnny's wife, ran the housing office and put up with rampant pretensions and prejudices. One day the wife of an engineer came in furious because she'd seen on a housing list that some Spanish-American family was assigned the apartment next to hers. It turned out the "Spanish-American" was Luis Alvarez, professor of physics at Berkeley and son of the famous Mayo Clinic doctor, Walter Alvarez.

The army brought in a bunch of college boys who had been drafted and had some scientific background to be lab technicians and help with calculations. They were called the SED or Special Engineering Detachment. And Van Vleck sent four juniors from Harvard who were in danger of being drafted. They were a remarkable bunch; I knew two them very well. One of them was Roy Glauber, who was a member of my group, and in afterlife became a distinguished professor at Harvard. The other one was Frederick de Hoffman, who even at that young age showed the talents as an operator which later took him to the presidency of General Atomics and then to the directorship of the Salk Institute. One day he came to me and said he had some questions about fission and we talked about it for a while. A couple of days later I met Richard Feynman in the hall and he said, "What have you been telling Freddie de Hoffman? He asked me some questions and when I answered them he said, 'but Serber said so-and-so.' " And another couple of days later Charlotte

Fig. 4.7　Luis W. Alvarez (1911–1988) received his B.S. from Chicago in 1932 and his Ph.D. in physics in 1936. He joined the University of California, Berkeley, in 1936. He was a staff member of the MIT radiation laboratory from 1940 to 1943, worked at the metallurgical laboratory in Chicago in 1943–44, and at Los Alamos in 1944–45. He went to Tinian in 1945 to help assemble the bomb for delivery, and was on board one of the two observation planes accompanying the *Enola Gay* to Hiroshima. After World War II he led the bubble chamber group at the Berkeley Radiation Laboratory, and was awarded the Nobel Prize in 1968 for his use of bubble chambers in discovering new particles. (Photo courtesy of the University of California, Berkeley)

said, "Well, I see you and Dick Feynman and Freddie de Hoffman have written a paper together." The paper was published as a Los Alamos report and after the war Freddie got it declassified and it appeared in the *Journal of Nuclear Energy* (38).

We found out that there were some advantages to being at an army post; the PX occasionally had unrationed goods. On the public address system at the laboratory there would be an announcement, "The PX has cheese today," and every secretary in the place would dash out to get her share. The army also had a theater that showed a movie twice a week. The lack of choice was partly alleviated by the fact they only charged fifteen cents. And the hospital was completely free. Perhaps that encouraged so many of our young population to have babies; it became known as "RFD," for rural free delivery. Groves once complained about it to Oppie, who replied that population control was not one of the duties of a scientific director. Kitty and Oppie were themselves guilty; their daughter Toni (formally Katherine, after her mother) was born in December 1944. I remember Charlotte and I on the grass outside the window of Kitty's hospital room congratulating Kitty, who was at the window, and Charlotte then relaying the latest gossip.

There was another aspect of Los Alamos life which was quite unfamiliar to us: the fact that we were supposed to work regular hours. Of course an experimentalist running an experiment might run all night, but a theorist had no excuses and it was eight to five, five days a week. We weren't allowed to take any papers home or even to talk about our work outside the technical area. This policy, while largely dictated by security concerns, I think had another effect as well: it encouraged an active social life on the mesa. There were lots of dinner parties, dorm dances, and weekends in the beautiful country around Los Alamos. There were various cliques: there were the hikers, the horsey set (to which Charlotte and I belonged, along with Robert and Jane Wilson and Hugh and Marjorie Bradner), the skiers in season and the fishermen, of whom Segrè was an avid member. One night Fermi was cross-examining Segrè to find out what the fascination of fishing was, and Segrè was explaining the fine points: you had to sneak up on a pool silently and not let your shadow fall on the water, and when you cast the fly it had to touch the water before the line. Finally Fermi said, "I understand. It's a battle of wits."

Another entertainment was the first fire in the residential area, which occurred in our house. I was away; I think I had a sore throat and they put me in the hospital overnight. In the back, between the two apartments, there was a furnace room with a wood-burning furnace. The Spanish-American janitor who took care of it used to try to pile in enough logs at night so that he wouldn't have to relight it in the morning. But that night he overdid it and sparks and flames were coming out of the chimney, the fire engine came, bell clanging, and all the neighbors rushed out to see what was going on. Bob Wilson wasn't one to miss such an opportunity. He went around to the other side of the house, got into our apartment, and appeared at the back door of our apartment carrying Charlotte in his arms, he in his bathrobe and she in her pajamas.

That janitor was also responsible for the start of Bob's career as a sculptor, because he used to leave his axe on our back porch. One evening after work Bob picked up the axe and a log and began chopping at it. Much to our surprise, after three or four nights of this it began to look like something. But the inevitable happened, and one night the janitor took Bob's log and put it on the fire.

Dinner at Edith Warner's became a Los Alamos custom. Miss Warner, a local resident, had a place at Otowi, just west of the bridge.[1] She could serve five couples at one large table, and half a dozen groups formed, each claiming the same night each week. Our group, the Friday nighters, suffered from the fact that all the men except Ed McMillan were called Bob: Bob Wilson, Bob Christy, Bob Davis, Bob Serber. We had to use last names.

We had another memorable evening, sometime in 1944, at an establishment we'd heard of called Swan Lake, somewhere up towards Taos, where one could make a reservation for dinner. Swan Lake turned out to be a large mansion. We—the Williamses, Wilsons, Serbers, and possibly another couple—were ushered into an expensive-looking drawing room and introduced to the chatelaine, who was beautiful, elegant, and got drunker and drunker as the evening progressed. She had drinks served, then excused herself to attend to dinner. Left to ourselves, we did a little exploring and soon found evidence of the difficulty of maintaining a chateau in the wilds of New Mexico. Charlotte and Jane called me upstairs to admire the bathroom: large, with gold-plated faucets and fixtures. But the remarkable thing was the floor, which looked like a perfect mirror until one stepped in, to discover it was a half-inch deep in water from a plumbing leak, held in by a one-inch sill. After some more drinks we were escorted to the dining room, paneled in dark wood with a fire in the fireplace, and were seated with our host and hostess at an antique Spanish table. As there were no other guests, we had already concluded that the restaurant was a tax deduction, not a business. Our hostess informed us she had a new French chef who'd just arrived that day and the meal was as good a French cuisine as one could expect at that locale. But the dinner was interrupted every now and then by an uproar in the kitchen, and the chef, in full regalia, would burst in and volubly complain to his mistress. She'd smooth him down, and as the evening wore on was more and more obviously making eyes at him, to the distress of her husband. We had brandy and coffee in the library, a large room lined on all four walls with books. We soon discovered a section of pornography and another of sports. Johnny Williams took down a volume and was riffling through the pages when he suddenly stopped and said, "By God, Serber, here's your picture!" It turned out to be the 1926 volume of sports records, and there I was with the rest of the Philadelphia Central High swimming team, National High School Champions. Altogether an incredible evening.

Late in December 1943, Niels Bohr and his son Aage arrived in Los Alamos bearing the aliases Nicholas and James Baker, and promptly became Uncle Nick and Jim. There was a story that just before coming to Los Alamos, in a Washington hotel, Bohr entered an elevator and found himself face to face with a woman he knew as the wife of the Austrian physicist Hans von Halban. He said, "Good evening, Mrs. von Halban." She answered, "I'm not Mrs. Von Halban now; I'm Mrs. Placzek. Good evening Professor Bohr." And Bohr replied, "I'm not Professor Bohr now; I'm Mr. Baker." Elsa Placzek's new husband was the Bohemian physicist George Placzek, who had visited the ranch one summer and was now a member of the British Mission to Los Alamos.

On the last day of December 1943, a secretary stuck her head in my door (the offices had no phones) and said Oppie would like me to come to his office. When I entered, Niels and Aage Bohr, Bethe, Teller, and Victor Weisskopf were already there. Oppie handed me a scrap of paper that looked as if it had been carelessly ripped from a note pad. It bore a sketch, and he asked me what I thought the sketch represented. After a minute I handed it back and said it looked like a heavy water moderated nuclear reactor. He then told me that Bohr had gotten it from Heisenberg. The question was whether it could be interpreted as a weapon. The Los Alamos experts gathered in that room all agreed it was useless as an explosive, and Bethe and Teller left to write a report to that effect for Groves, with whom Bohr, himself no expert, had previously raised the question. On New Year's Day, Oppie wrote Groves an account of our meeting and enclosed the Bethe-Teller memorandum.

One morning in January 1944, Charlotte came to my office with a wire from Mary Ellen Washburn in Berkeley, saying that Jean Tatlock had committed suicide the night before. She asked me to break the news to Oppie. When I got to his office I saw by his face that he had already heard. He was deeply grieved.

For the first year all the experimental groups—Bob Wilson and the cyclotron, John Manley and the Cockcroft-Walton, and Johnny Williams on the Van de Graaff—were busy measuring cross sections, the chemists and metallurgists were getting their techniques in order, Ed McMillan was firing guns and testing assemblies down in a canyon, and Seth Neddermeyer was firing shots and finding out how difficult implosion really was. The theorists were developing ideas about implosion. In the fall of 1943 Johnny von Neumann came around and talked about shaped charges and the Monroe effect, which weren't directly applicable but put in everybody's mind the notion of how metals flow plastically under the pressures generated by detonation waves. James L. Tuck had invented explosive lenses to shape the waves.

Gun assembly was still the favored method because it seemed so much easier than implosion. That is, until the summer of 1944, when the first reactor-produced plutonium began to arrive in Los Alamos and was found to be poisoned by the isotope of plutonium ^{240}Pu. Plutonium-239 was the isotope we wanted, and ^{240}Pu was made in a secondary reaction when the ^{239}Pu absorbed a neutron. The trouble with ^{240}Pu was that it had a very high spontaneous fission rate and gave a big stray neutron background. So for plutonium the gun assembly became impractical; one couldn't do it fast enough to beat the stray neutrons. The whole laboratory was reorganized with the idea of developing implosion.

My group was left in charge of the design of the uranium bomb with the gun assembly—the Hiroshima bomb. This, by the way, was used without any test at all. It was a comparatively straightforward job. There was only one unanswered problem: If you're trying to assemble three critical masses' worth of uranium and want to break them up and separate them into two pieces—a target and projectile—the question is, are the two pieces supercritical or subcritical, able to sustain a chain reaction or not? If you made them into spheres and surrounded them with reflector, they'd each be one and a half critical masses. Is the actual configuration super or sub, yes or no? You'd better know or you'll blow yourself up. With such odd shapes it was impossible to make a calculation, but I thought of a way of answering the question. I had Bob Wilson make scale models of the pieces, where ^{235}U was represented by graphite and the reflector was represented by graphite with some absorber added. He could irradiate each of the models with a burst of neutrons from the cyclotron and measure the rate of decay of the slow neutron density in the model. Slow neutrons gave a convenient time scale for measurement. He also made a graphite sphere which was calculated to be the model of a one critical mass sphere. All one had to do was compare the decay rate of the model with that of the sphere. If the model decayed faster, it was subcritical and you were safe. You could also bring the two pieces together gradually to see at what point the system became critical and how supercritical it became when the pieces came together, information that could be used to calculate predetonation probability and efficiency.

The other thing my group was responsible for was the interpretation of integral experiments. Integral experiments were the ones that were done after enough ^{235}U or ^{239}Pu began to arrive from Oak Ridge and Hanford in quantities you could see and make into little spheres. You could put a neutron in and see what happened to it. You didn't have to calculate from scratch, from the fundamental cross sections, to see what would happen; you could see what did. Of course, it was also interesting to see if the calculations were right. If you put a neutron source in the center of a sphere, more neutrons would come out than were emitted by the source, and as more and more material came in and you could make your spheres bigger and bigger, the multiplication of neutrons got bigger and bigger too. You could extrapolate to where the multiplication would become infinite, and that was the critical size. Of course it got quite exciting in the very

Fig. 4.8 Isidore Isaac Rabi (1898–1988) was born in the Austro-Hungarian empire in 1898 but moved with his family to the United States the following year. He obtained his B.A. from Cornell in 1919 and his Ph.D. from Columbia in 1927. He joined Columbia as a faculty member in 1929 and remained associated with Columbia for the rest of his life. He helped to build up Columbia's Physics Department into one of the finest in the country. During World War II he was an assistant director of the MIT Radlab and visited Los Alamos several times. He played an instrumental role in the founding of Brookhaven National Laboratory in 1946–47. He received the Nobel Prize in Physics in 1944 for the resonance method of recording the magnetic properties of atomic nuclei. He was a member of the General Advisory Committee of the Atomic Energy Commission and succeeded Oppenheimer as its chair from 1952 to 1956. (Photo courtesy of Columbia University)

end when one was adding the last little bit. Another method was to give the sphere a burst of neutrons from Bob Wilson's cyclotron and see how long it took the neutron density in the sphere to decay.

But before nearly pure ^{235}U was available, there was an earlier integral experiment, using the uranium enriched to 14 percent ^{235}U put out by the first stage of Ernest's electromagnetic separators at Oak Ridge. This was the water boiler, a small, low-power reactor using the enriched uranium dissolved as a salt in ordinary water. The leader of the water boiler group was Don Kerst, with whom I'd worked on the betatron (later Don brought the 15 MeV betatron from the University of Illinois to Los Alamos for use as an X-ray source), so naturally he asked me for help. I made some preliminary esti-mates for Don, but then Bethe asked me to turn the project over to Bob Christy. On May 9, 1944, word went around that the water boiler was ready to try out. A bunch of us went down to the lab in Los Alamos Canyon. Don was sitting in the chair in front of the control panel, but at the last moment he got up and offered the chair to Fermi. Enrico sat down and turned the knob that pulled out the con-trol rods, the counters clicked faster and faster, and the spot on the oscilloscope climbed higher and higher till it disappeared at the top of the screen. Enrico pushed the control rods back in.

I didn't have much to do with the implosion development. I only contributed one invention, the RaLa method of measuring implosion (RaLa was an abbreviation for radiolanthanum). You would take a radiolanthanum source of gamma rays and put it in the center of a sphere to be imploded, and outside you would put detectors to mea-sure the intensity of the gamma rays. When you set off the high explosive, if the sphere were compressed the intensity would drop. By how much it dropped you would know how much compression there was. Bruno Rossi and his group developed the technique and carried out a number of such experiments.

On the day Paris was liberated, in August of 1944, Rabi happened to be around, and he and Viki Weisskopf decided the event wasn't being properly celebrated. They marched around the residential area bellowing the Marseillaise, inviting everyone to join the parade and it turned into a sort of holiday.

On April 12, 1945, the news of the death of Franklin Roosevelt struck us like a blow. Oppie scheduled a memorial service for the fol-lowing Sunday where he spoke briefly, eloquently, and very movingly.

May 8, 1945, was VE Day, and the war in Europe ended without a German A-bomb. In the British *Farm Hall Report*—secretly recorded conversations between captured German scientists after the war— Heisenberg told Otto Hahn that he had never bothered to calculate the critical mass of a uranium bomb because he didn't see the point; he didn't think separating any considerable amount of ^{235}U was a practical proposition. As Heisenberg had indicated to Bohr, the Germans were working on a heavy water reactor, and Carl F. von Weizsäcker had pointed out the plutonium route to a bomb. But to get enough heavy water to build production reactors also seemed impractical. The German experimental physicist Walther Bothe had badly misled his colleagues by reporting that graphite was useless as a moderator for a reactor, not realizing that ordinary commercial graphite was contaminated with boron, a strong neutron absorber.

But the end of the war in Europe had no effect on the pace of work at Los Alamos, which was nearing its climax; a desperate and bloody war was still going on in the Pacific.

During the spring of 1945, Oppie was involved with the considerations of the Interim Committee, a body set up by President Truman to advise him on the use of the bomb. It was chaired by Secretary of War Henry Lewis Stimson, and Secretary of State designee James Francis Byrnes and Gen. George C. Marshall were members. Oppie, Fermi, Compton, and Lawrence were on the Scientific Panel of the committee. At Los Alamos, Oppie discussed with me the problems facing the Interim Committee. He told me about their briefings, about the plans for an invasion of Japan in the fall, and that the Medical Corps of the armed services had been told to prepare for half a million casualties. Given this background, we had no doubts about the necessity of using the bomb. We spoke of it as a "psychological weapon," and were sure dropping a bomb on a Japanese city would end the war.

During this period a number of the scientists at the Metallurgical Lab in Chicago advocated a demonstration of an atomic bomb to the Japanese before deciding on using it. Szilard asked the Los Alamos scientists to join a petition urging this course. There was a meeting one evening in Bob Wilson's lab to consider the question. I remember that Oppie spoke, but no details of the discussion. However, the outcome was that there was no Los Alamos petition. I think there was very little support for the idea in Los Alamos. I was against it: the psychological effect of a demonstration would be very different from that of an actual use. In July the Scientific Panel of the Interim Committee reported its conclusion that a demonstration was unlikely to end the war and said it saw no alternative to direct use.

That brings us to the Trinity test on July 16, 1945. Oppie played a dirty trick. He sent out a memorandum saying all the members of the Coordinating Committee who didn't have duties at Trinity would be bussed to a viewing site about twenty miles from the explosion. The Coordinating Committee, which considered technical matters, consisted of all the group leaders except one—Charlotte, the only woman group leader. Of course, she was outraged by this piece of sexism, but Oppie wouldn't back down: the reason he gave was that there were no sanitary facilities at the viewing site. On the evening before setting out, I ran into Edward Teller on the street and he remarked that Oppie's memorandum had said to watch out for rattlesnakes and he said, "What are you going to do about rattlesnakes?" I said, "Well, I'll take a bottle of whiskey." Then he brought up the notion that the atmosphere might be set on fire by the bomb and he said. "What do you think of that?" and I said, "I'll take another bottle of whiskey." In Richard Rhodes's book *The Making of the Atomic Bomb*, he seemed to think that meant I was fearful, but it wasn't so. It was just a wisecrack.

We left for the viewing site that night and got there about three in the morning. It was cold, damp, and there were thunderstorms around which delayed the test. At about 5:30 it cleared up a bit and a warning rocket went off in the distance. We all lay down facing the test site. We'd each been issued a piece of welder's glass to shield our eyes, and were supposed to hold them up while awaiting the explosion. Of course, just at the moment my arm got tired and I lowered the glass for a second, the bomb went off. I was completely blinded by the flash, and when I finally began to regain my vision the first thing I saw was a violet column that must have been very bright and was thousands of feet high. In about half a minute my vision cleared and I saw a white cloud rising—to what must have been twenty, thirty, forty thousand feet high. I could feel the heat on my face a full twenty miles away. The fireball was about as bright as the sun on a clear summer afternoon. After about a minute and a quarter, I heard the crash of the explosion, which was like very loud thunder and which reverberated for several seconds around the surrounding

Fig. 4.9 The Trinity blast at thirteen seconds. (Photo courtesy Los Alamos National Laboratory)

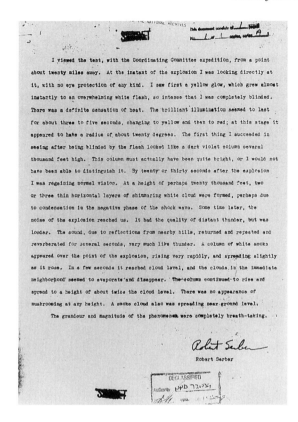

Fig. 4.10 My report on the Trinity blast.

hills. Don Kerst hadn't come along because he'd been sick the day before. He got up and watched from Los Alamos, about two hundred and fifty miles away. He saw the flash, and about twenty minutes later heard the explosion.

The next day we each were asked to write an account of what we had seen. The text of my report is given in figure 4.10.

There had been a pool in Los Alamos to guess how big the explosion would be, and I think I guessed around 12,000 tons' equivalent of TNT. Rabi, who had guessed 18,000 tons, won. Almost everybody else had guessed a lot lower, except Edward Teller who guessed very high—whether out of optimism or gamesmanship I don't know. Afterwards, Rab told me that the way he did it was to go around to the various theoretical groups and ask what they'd calculated the explosion should be, and bet on that. As Rab expected, all the Los Alamosites were so impressed by the difficulties of the implosion technique that they had guessed too low. In 1959 Charlotte and I were in London and heard that Rab had been taken to a hospital there after suffering a heart attack. We went to see him, and at the moment he was practically having another heart attack because he'd just been reading Robert Jungk's book *Brighter Than a Thousand Suns*, the first popular book on the atomic bomb, and Jungk had written that I had won the pool, not Rabi. Rab was proud of having won.

I don't recall any celebration at Los Alamos when we got back from Trinity. I think we were too tired that night and just went to bed.

FIVE

Tinian, 1945

The Trinity test was on a Monday, July 16, 1945. That same day the cruiser *Indianapolis* left San Francisco with the Hiroshima bomb—^{235}U, gun assembly—en route to deliver it to the Marianas Islands in the Pacific war zone. The following Friday I left Los Alamos for the same destination with what was called Project A, the group that was to assemble the bombs for delivery to Japan. Norman Ramsey was the group leader; other members were Luis Alvarez, Phil Morrison, Bernie Waldman, Larry Langer, Harold Agnew, Charlie Baker, and my old friend Bill Penney.

It was the first time I was away from Charlotte for any length of time, and I wrote her every few days. The letters are reproduced below, with added comments and explanations in bracketed italics. Our letters had to pass Army censors, so some things had to be omitted which I can now add. It was forbidden to describe one's travels—route and destination. These restrictions were (presumably) lifted when peace was declared, and in a letter from Tinian dated September 2, 1945, the day the Peace Treaty was signed in Tokyo Bay, I went back and described our trip to Tinian. Here is the excerpt dealing with that trip.

Sept. 2 [*Tinian*]. . . .

Now that the war is over I don't see why it isn't ethical to back track and tell you a little more about our trip out here (since this is going private post, & not through any censor). [*I meant to ask Luis Alvarez, who was returning to Los Alamos, to deliver the letter.*]

You remember that bus that took us out on Friday morning. We went right out to the Albuquerque airport, had lunch there, then moved over to Kirtland field where an OD painted C-46 was waiting. This is a fat, four engine job—it said on its side "Troop Carrier Command." The rumour was that our first stop would be Las Vegas. I sure enjoyed that first ride. Aside from the fact that I always like planes, the atmosphere was exciting—the bucket seats down both sides in which you sit on your parachute, the signs stenciled on the walls "155 mm howitzer or 4 x 4 truck front wheels here"—so much more matter of fact and businesslike than the commercial planes.

We flew west. It was hot as hell in the plane waiting for the takeoff, but nice and cool after we got up. We left the Rio Grande Valley behind and started across the New Mexico–Arizona badlands. In no time at all the San Francisco mountains loomed ahead. We turned north and crossed the Grand Canyon at just about the point we first met it that evening on our trip with Billy. I thought we'd be in Las Vegas in a few minutes, but when I crossed over and looked out the window on the other side of the plane, there was a view below that couldn't be anything but Zion Canyon. We kept heading north, and about 2 $\frac{1}{2}$ hours after leaving Albuquerque we landed at W-47, which turned out to be our old friend Wendover, on the edge of the Salt Lake desert. Wendover doesn't look like it used to. It's just as bad, but much bigger. [*Wendover was where the 509th Composite Group, the Air Force group that was to drop the bomb, was stationed*.] I've already written of our doings there, so I'll skip to Tuesday morning.

The letter about our doings in Wendover:

Sunday, July 22 [*Wendover*]

Dearest Charlotte:

The hard part of writing is not going to be finding things to say, but in figuring out what will get by.

So far everything has gone quite smoothly. It seems well organized, and none of the predicted snafus. The plane ride from Kirtland was very fine. I really enjoy flying—in this I seem almost unique. Everybody else either sleeps or reads detective stories, but I can't keep away from the windows. The scenery on this trip was well worth describing, but the description would be too much of a giveaway. Anyway, we arrived at W 47, got our barracks (which are dirty, but otherwise comfortable), had dinner at the officer's club, which is quite a nice joint. We spent the evening by Luis teaching me to fly a Link Trainer, which, if you don't know, is an airplane that stays on the ground.

Fig. 5.1 The work of an army photographer, who must have taken wedding pictures as a civilian. This photograph of me was taken at Wendover Air Force Base, where the 509th Composite Group, the air force unit that was to drop the bomb, had its headquarters.

The next day was spent mostly in getting processed, which is a fairly simple procedure, but involves some waiting around. All the papers were OK. The special passport which says, "The bearer is a government official proceeding abroad on official business," is quite impressive. It would make a nice souvenir, but I'm afraid it will have to be turned back. Also the orders and permits from the Joint Chiefs of Staff. We got our AGO cards. My assimilated rank is Colonel all right. Luis was disappointed, he's a Lt. Colonel. Most of the others are Captains. I think I was the only one that didn't need some extra shots. We got dog tags, and all the issue junk that was on the list, which almost completely fills a good sized duffle bag. The Service-Pack bag was a very fine idea. Everybody else is trying to get all their stuff in the duffle, and having a very hard time. It was also a good idea to get everything in advance. The clothes available here were all the simple GI stuff. No shoulder patches, no brass buckles, no officer's cap, etc. Last night I got into uniform—I guess two years of ROTC left a secret mark, because I was told I looked natural and natty. I don't know about that, but the others didn't look too soldierly to me. I at least fooled some GI's into saluting, and managed to get out and into the

Fig. 5.2   Henry and Shirley Barnett at Los Alamos.

post without a pass (this wasn't just a trial—I didn't have a pass). The occasion for leaving was to visit the hovel Penny has to call home. I had dinner at the Officer's Club with Sam and Pen, and then went out to their place. Penny is a brave girl to try and survive here, but she seems in reasonably good spirits, since she is seeing Sam for the first time since the War. I don't know how long she'll be able to stick it though. [*Penny was the wife of Sam Simmons, a young physicist recruited by Los Alamos in June 1945 from the MIT Radiation Lab, where he'd been doing liaison with the Air Force. The security people seem to have gone bonkers in Sam's case. He wasn't told where he was going. He was given a train ticket to Chicago and told that at the Chicago station a man with a white carnation in his buttonhole would approach him with further orders. It really worked; there was a man with a white carnation who gave him another train ticket to Lamy, New Mexico, and an envelope containing instructions to find his way from Lamy to Santa Fe and report to 10 E. Palace. This was a slight error. Los Alamos' Santa Fe office, run by Dorothy McKibben, was actually at 10 1/2 E. Palace, set back from the street down a path next to number 10, which was a restaurant. Sam was surprised when he got there but after his*

Chicago experience he could believe anything, so he went in and ordered a bowl of soup. After half an hour with no contact made he began to fidget. Finally the counterman, inured to the idiosyncracies of his neighbor, came over and said, "Say buddy, I think you want next door." Sam didn't stay long at Los Alamos, He was sent to Wendover, in charge of the tests there.]

It's now about 10 o'clock Sunday morning, and my immediate job is to pack all my stuff so it can be weighed in this afternoon. I forgot to tell you that I spent most of yesterday afternoon working on the cigarette angle, and managed finally to get half a dozen packs. Tell Henry the lighter he lent me is working beautifully, and has been much admired, even by the Air Corps. [*Henry Barnett was a captain in the Army Medical Corps and an important man at Los Alamos—the pediatrician.*]

I can think of only one thing I (or rather you) forgot: some container for shipping clothes back from here. However there will undoubtably be room in some of the suitcases being sent back. I'll see if I can get my stuff into Luis's.

You'll get this just about on your birthday, so the best kinds of greetings and wishes and lots of love. I hope you'll like the present.

Are you and Shirley batching it together now? How is the pooch getting along? I'll write again soon, but it may take some time for the letter to arrive so don't worry about it. [*Shirley Barnett was Henry's wife and assistant executive secretary in Oppie's office. Re the pooch: I think Charlotte got a dog while I was away, but I don't remember it.*]

With much love, Bob

Returning now to the September 2 letter from Tinian (at the point "I've already written of our doings there, so I'll skip to Tuesday morning"):

We took off fairly early in the morning in a C-54, a huge beautiful silvery 4 engine transport plane. This was one of the "Green Hornets," our private transport planes, which carried project staff between Wendover and Tinian. It had a green stripe down the sides, green wings forward, and the insignia was a winged calf hopping from one palm studded island to another. These planes were unusual, in not fitting into the regular ATC system, and there was always a lot of curiosity and comment about them wherever we went. There was a lot of freight lashed to the floor, and canvas seats stretching along both sides of the fuselage.

We crossed the Sierras, flying over Tahoe, Sacramento, and landed at Hamilton Field, near San Raphael, in Marin County. A good clean hot California day. That must have been about 10:00 am. We had to fill out assorted forms and see a movie about "ditching" (i.e., landing at sea). We had all afternoon to wander around the field and post, & work on the problem of getting a cigarette ration card, which we did successfully. It was a little chilly crossing the Sierras, so I bought a field jacket at Hamilton. Late in the afternoon Luis and I were standing on the field when we saw a plane come over at a terrific speed—about 500 miles/hr, trailing smoke so we thought he was going to crash. But he buzzed the field and landed—it was a P 59, a jet plane, very low slung compared to conventional planes. It created quite a furor. Hundreds of people rushed out to see it—the first jet job seen at Hamilton.

The terminal at Hamilton Field was really something to see—the main center of air transport to the Pacific. Terrifically busy, and all kinds of army, from the real brass to extremely tough looking fighting squads, loaded down with side arms, knives and blackjacks.

The trip from Hamilton to Oahu couldn't have been better planned for scenic effects. We left San Francisco at sunset, out over the gate, and then there was a full moon to light us on our way. It was a wonderful night. We flew high and way below us, just off the surface of the sea, the big cloud banks drifted by. I sat up most of the night watching it. We passed over a couple of big convoys—you could see the ships as the moon's wake passed over them. For 50c invested in Hamilton we had had in-flight lunch put on board, so along about 2 am I had sandwiches. There was one minor tragedy—both thermos were full of soup, and there was no coffee. Finally I stretched out on one of the canvas seats and got a couple hours of sleep. It was pretty cold & the field jacket came in handy. There was lots of room, only nine of us in the plane. Most of them preferred to sleep on the floor.

We flew in over Diamond Head at sunrise (a 2,300 mile hop). It's an almost perfect volcano rising from the sea, with a round crater on top. Behind lay Waikiki, Honolulu, Pearl Harbour. We landed at Hickam Field, which took such a beating on Dec. 7. The new passenger terminal is really beautiful—an open wooden structure with a big patio down the center—looking like a Hollywood set. The palms and flowers are magnificent. Huge flowers, deep red, 6" across.

We had breakfast and then were taken to the VOQ—Visiting Officers Quarters. These were the huge, pre–Pearl Harbor barracks, and showed it. Bullet holes, large hunks knocked off by bombs, patched up places. It has taken a hell of a beating. Here we had showers—we didn't know it then, but that was the last hot water we were to see.

Then the annoying part: no one would tell us when our Green

Hornet would leave again, so we had to hang around, when we were dying to see Honolulu or take the round the island bus. Not being on hand, leaving the VOQ without permission, can be treated as desertion, since flying the Pacific is classified as hazardous service. However about 10 o'clock we were told we were free until 2:00. So Luis and Larry and I hopped a bus into Honolulu. The bus passed through a big section with a Chinatown look, and very squalid. Most of the stores had Jap names on them. Even the center of town looked pretty sad—no civilian life left to amount to anything, just a huge military camp. We went on to Waikiki, which still has some of the attributes of a swank resort, though the Royal Hawaiian is now a naval hospital. I can easily believe that Waikiki is the finest beach in the world. Or would be if you knocked down all the hotels and clubs encroaching on the beach and barring off each its little patch. The water is that brilliant clear turquoise to deep blue again. The surf is wonderful—row after row of huge breakers coming in for a mile or more. The guys who ride the surfboards in are really good—just like the movies, and lots of them. We sat in the courtyard of the Moana Hotel, which faces the beach, and watched them awhile. The beach was crowded. It looked a little odd though—hundreds of guys and no girls. There's a huge banyan tree in the Moana courtyard. The only thing in the tree line that is worth mentioning in the same sentence is a giant redwood. Then we had lunch at the Moana, which seemed to be the only decent place left, to judge by the way it was mobbed. After which we returned to Hickham and killed time. About 8:30 in the evening we got a call to report to the field. While waiting there we got coffee and doughnuts from the Red Cross, but what was really notable was the big bowls of fresh pineapples they had sitting around.

Our elegance was slightly shorn on the rest of the trip—they put 14 extra passengers in with us: a couple of officers, a radio technician, a crew of GI's equipped for fighting. That was because the next hop was not so long. It could be shortened by stopping at Johnston Island, 600 miles out.

We had a lovely view of the lights of Honolulu and Diamond Head in moonlight, then west over the Pacific. We reached Johnston about 1 am. It's just big enough for the airstrip. In fact Johnston Island is 5,000' long, and the airstrip is 7,000.' They built it over the reef. Griddle cakes and coffee at Johnston, then we took off again for Quajalein [*Kwajalein*] in the Marshalls. This is the world's biggest atoll. It took a terrific pounding when we took it from the Japs. The palms look very sad, stripped to the trunk.

What with the wake of war and the usual unsightliness of the army,

Fig. 5.3   Tinian Island, the north end of which had been converted into one huge airfield.

Quajalein was a pretty sad place. We stayed only long enough for breakfast, and hopped off for the Marianas.

On the night-trip from Oahu to Quajalein I came within an ace of missing your birthday completely. We left on Tuesday evening and arrived Thursday morning. The skipped day was the 25th though. [*This is an error. We left Wendover on Tuesday the twenty-fourth and spent the twenty-fifth in Hawaii, so it was indeed the twenty-sixth—Charlotte's birthday—that we missed. Not anticipating this, I had left a present to be delivered to her.*]

Quajalein is only 5° above the equator. Shortly after leaving I noticed that the sun was shining in the right hand windows, although we were going west. It was quite confusing, for a while, to get used to the fact that at noon the sun is in the north.

We passed a number of atolls, which are fine things to see from the air. We traveled around and through a lot of towering cumulus, ducking thunderstorms, and, after about 1,300 miles, set down at Saipan. It was our first advance base field, and impressive, with rows and rows of B29's lined up along the taxistrips, and piles of junked planes off in the corners.

It's a pretty island, all green, with a 1,500' mountain looking like just the spot for a country villa. About 10 or 15 miles across. We could see the channel, 5 miles or so wide, and on the other side Tinian, flatter and smaller than Saipan. On most of the coast the cliffs fall off 50' to the sea, with a creamy line of breaker at the base.

At Saipan we dumped the extra passengers, and had a medical examination, which consisted in a very bored enlisted man giving a very quick glance into your throat. Then, after waiting around a couple of hours, we hopped off again for Tinian. This was our shortest jump—7 minutes.

If we thought the airfield on Saipan was big, we soon learned better. North Field on Tinian is the biggest in the world. Four parallel 8,000' runways, with parking strips between, and the 29's lined up by hundreds. It's incredible to see, and to think of such a base existing so far from home.

There were some familiar faces waiting for us—notably Jim Nolan's. Our camp is just off the edge of North Field, and arriving at camp ends the story of the trip.

Jim Nolan was the Los Alamos obstetrician, also a captain in the Army Medical Corps. Jim had flown over carrying the plutonium for the Fat Man, the implosion bomb, in a little handheld case. It seems odd that General Groves sent the Los Alamos obstetrician and pediatrician to the Pacific but not our radiologist, Louis Hempleman. I was told it was because Louis was a civilian.

The cruiser *Indianapolis* had delivered Little Boy, the gun assembly bomb, to Tinian on Charlotte's missed birthday, the day before we arrived. She left en route to Leyte in the Philippines, whence she was scheduled to take part in the invasion of Japan in the fall. On July 29 she was torpedoed in the Philippine Sea (a famous story reprised, among other places, in *Jaws*). Leyte never noticed the *Indianapolis* was overdue, and it was four days until a Navy plane happened to spot the survivors; 300 were rescued of a crew of 1,200.

While we were at Tinian we saw several big strikes go out against Japan—hundreds of planes, one right behind the next. The runways ended at the top of a fifty-foot cliff above the sea, and the planes were so heavily loaded they'd go down to the end of the runway, barely take off, and disappear below the level of the cliff. Ten seconds later you'd see them coming up again—which was a little exciting because, we were told, they were loaded to the point where one percent of them were expected to crash into the sea on takeoff. Gen. Curtis LeMay figured his losses were less that way than flying more planes to Japan and back.

At our camp two Quonset huts served as our laboratories where the bombs were to be assembled. One was for Fat Man, the other for Little Boy. At Los Alamos I had given the bombs names descriptive of their shapes, the gun assembly was the Thin Man, taken from the title of the Dashiell Hammett detective novel, which had recently been made into the William Powell–Myrna Loy movie. The name the Fat Man for the implosion bomb then followed naturally, after Sidney Greenstreet's role in *The Maltese Falcon*. The original Thin Man bomb, designed to contain the high-velocity gun intended for plutonium assembly, was so long it filled two bomb bays of a B-29. In drop tests the Air Force ran into trouble getting the two hooks to release simultaneously and was relieved when a much shorter, one bomb bay version was substituted, sufficient for the low-velocity uranium assembly gun. They named the new version Little Boy, by comparison with Thin Man.

Fri. July 27 [*Tinian*]

Dearest Baby:

There's a tremendous amount to tell, but all the most interesting parts will have to wait.

The trip here was wonderful. At this end everything seems to be well organized and going OK.

The privileges of a Field Grade rank consist primarily in living two in a tent instead of twelve or so in a Quonset hut. Luis and I share a tent, which, of course, is not quite completed, and which gets worked on a little every day. For 25c/wk an EM polices it, makes the bed, etc. The food is pretty good, cigarettes are 5c a pack and whiskey 95c a quart.

The climate is a little warm, and very very humid. One just gets used to being soaking wet in the day, and good and damp at night. But at that it's not much worse than Phila in summer.

Also the vegetation is rather strange, if you happen to notice. It just manages to be different enough to be confusing. Nothing you could call a dandelion though.

The hours people work are adjusted somewhat to climate. Breakfast before seven. Quit at eleven. Start again at 13:00 (pardon me, 1:00) till about 4:30.

Last night we saw a show, along with about 3,000 other guys. It was "This is the Army," and it was very good, and a great success. Tonight, there was another, with Eddie Bracken, but we didn't see it. (These aren't movies, but the real thing.) If it goes on like this, it's a better season than you'd get in New York. The guys really seem to get an enormous kick out of it. [*I soon discovered the advantage of attending these performances in company with a general or admiral. Their staff would occupy choice seats which they would turn over when we arrived.*]

I hear that Dick [*Richard Feynman*] isn't coming after all. It's a shame because he was so eager to do it. However there doesn't seem much point to both of us being here.

Jim arrived the same day we did. He enjoyed his trip over, and has been spending the time since constructing furniture for his tent.

Its sort of dismal writing and having to leave out all the important things. But I'll tell you about it later.

Lots of love, Bob

Tuesday, July 31 [*Tinian*]

Dearest:

The form of life out here is quickly taking shape. We get up about 6:00 am, and have breakfast at 7:00. Then everybody goes to work until 11:00. Lunch then, and lie around in the sun (if it's out) till 1:00. Work till about 4:30, dinner about 5:00, kill time until 7:15 when there's a movie or a show. The movies are preceded by 15 minutes of news and combat reports. After the movies to the officer's club, for a drink or a beer, or a coke, and to bed by about 10:00.

Fortunately my principal job hasn't materialized—no trouble. However I'm kept very busy answering all sorts of questions and making all kinds of calculations. Norman and Jim are my best customers.

The weather reminds me very much of Phila in a summer heat spell. A lot of rain. The night before last it was practically a hurricane, but our tent weathered it in fine shape. Some others weren't so fortunate—the clever boys who had taken off some of the flaps to get more air. All yesterday and today it remained cloudy. Henry's raincoat is getting a lot of use, and so, in spite of previous doubts, is my helmet, which turns out to be a good rainhat.

By the way, all my equipment is fine (except for lack of buttons on the shirts). I got a couple of pairs of shorts on the way, which are very good for the sunny days. There is no uniform on this island; everyone dresses according to their whim. The only formal restriction is a sign in the Officers Club which demands that pants be worn after 17:00. I am lacking a laundry bag. I traveled down to the other end of the

Fig. 5.4 Colonel Tibbets, who piloted the plane that dropped the bomb over Hiroshima, was naturally concerned that his plane might not survive the blast. He drew a picture of the maneuver he planned to execute and asked me what would happen to him.

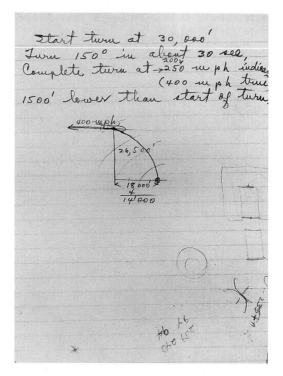

island to get one this morning, and of course discovered that the Quartermaster's Store was closed for inventory. Just like home.

Peer and Frances got in a couple of days ago. They gave Peer the second oldest plane in the transport service, and it had to turn back because an engine conked out. [*Lt. Col. Peer de Silva was an Army Intelligence officer at Los Alamos; Commander A. F. (Frances) Birch was an engineer who, along with Ed McMillan, was responsible for gun design and testing.*]

We went swimming one afternoon. The best beaches were closed by rough water—the currents around here seem quite treacherous, and they have lifeguards who are very strict about what you can't do. So we went to what is called the Yellow Beach. It isn't very good for swimming, because they won't let you past the reef, and inside the reef it's nowhere more than waist deep. But when you go in you discover that it is magnificent, in a way no beach or no swimming we've ever seen before is. The water is crystal clear, and full of little bright colored blue fishes, with some striped affairs swimming around too. And the bottom is all little ravines and canyons; coral, in bright colors, like fantastic vegetation; brilliant rocks in red, green, pink, yellow. The rocks are very jagged and sharp. You have to swim very carefully under water through narrow passes, and follow the valley floors, which are bottomed with clean yellow sand. Under this clear water you can follow the topography for 50 feet.

As you can gather, I liked it. The landlubbers I was with didn't, because they tried to walk around and kept cutting themselves on the rocks.

Another thing we have good here are the sunsets, which go in for pink and greens in a big way.

There's a lot of sugar cane growing around. Most of the other stuff which grows so lushly is very pretty and delicate, or very brazen, or very jungly (like out of a set for the Emperor Jones). However nobody seems to know the names of any of it.

One thing that's surprising is how little drinking there is of anything but beer and coke. And the only gambling is one crap game at the Officer's Club, which is for stakes far too high for any of our own guys.

No mail has arrived here from the Site yet.

Thus ends, for the moment, our account of life on_____.

Lots of love, Bob

Another job came up for a theorist. Col. Paul Tibbets, the commander of the 509th, was to fly the plane that dropped the bomb and was naturally concerned that his plane might not survive the blast. He drew me a little picture and wrote a little description of the maneuver he planned to execute and wanted to know what would happen to him (see figure 5.4). His description read:

Start turn at 30,000'
Turn 150° in about 30 secs.
Complete turn at [arrow] 200–250 mph indicated (400 mph true)
1500' lower than start of turn.

I did some calculating and wrote him an answer (see figure 5.5):

Shock reaches plane at distance of 10.6 miles
Pressure in shock wave is 0.16 p.s.i.
Time after explosion is 41.4 sec
    "   "   drop   " 84.7 sec

I didn't know much about planes, but I assured him that he would be perfectly safe.

On the night of August 6th we attended the briefing of the crews

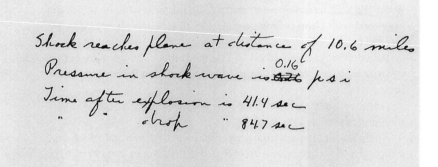

Shock reaches plane at distance of 10.6 miles
Pressure in shock wave is~~that~~ 0.16 psi
Time after explosion is 41.4 sec
   "   "   drop   "   84.7 sec

Fig. 5.5   Here is my reply to Colonel Tibbets. I assured him his plane would be safe.

who were to fly the first mission. The operations order listed the planes and their crews. Underneath it said, "Breakfast at 2 AM, Takeoff at 3 AM" and then, "Bombs: Special." There were three planes going. There was Tibbets's plane to carry the bomb with Capt. William S. ("Deke") Parsons to arm it. The second plane carried an aluminum cylinder containing a pressure-measuring apparatus which had been Luis Alvarez's work, and Luis and Harold Agnew flew on that plane. The third plane carried a Fastax camera to photograph the explosion, a very high-speed camera which could take up to 8,000 frames per second; Bernie Waldman and Larry Johnston (I think) flew on that one. During the takeoff Phil Morrison and I went to the headquarters of the commanding general on the island. We were supposed to be prepared to give advice about what to do if the plane carrying the bomb crashed on takeoff.

The next morning we got the news the mission was a success. Of course, we were overjoyed. We were radioed back the readings from the pressure apparatus that had been dropped with the bomb. Bill Penney and I were sitting in one of the Quonset huts, calculating how big the explosion had been, and we almost had the answer, just on the verge of getting it, when, on the Armed Services Radio, we heard President Truman announce that it had been 20,000 tons—a slight exaggeration but not misleading.

Aug. 7 [*Tinian*]

Dear baby:
    Well we seem to have gotten a pretty good press, despite your mis-

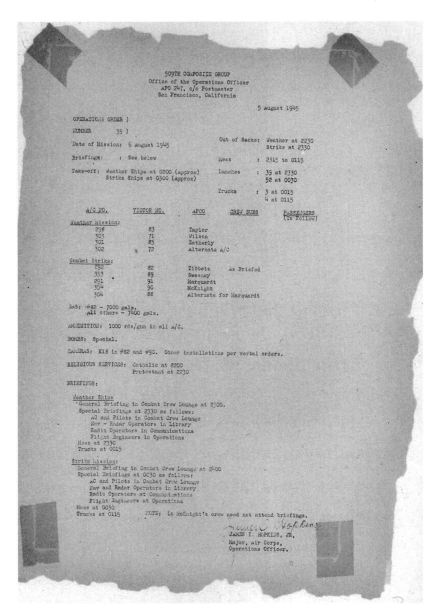

Fig. 5.6 The operations order for the Hiroshima mission, listing the planes and crews. Note "BOMBS: Special."

Fig. 5.7 Working at
Tinian. (Photo courtesy
of Harold Agnew)

givings. Here everything went like clockwork—no headaches and just as expected. The place is all steamed up as you can well imagine. Our boys (not the civilians, but the army) have been taking a lot of kidding from other units, so they are on top of the world right how. Yesterday afternoon the Group had a beer & hot-dog party on the ball field, with a short and good speech from the CO. We had a few days with not too much to do, and managed to go swimming every afternoon (here there are no restrictions on using government cars for anything you please). Then it got hectic again, and we've been rushing around pretty madly, but still get in at least a dip in the ocean.

The ocean continues to be marvelous, the sunsets gorgeous, the food OK, and life very exciting.

I hope we may do some good in the direction of shortening the war. I haven't had time to pay much attention to the radio today. What was the reaction like?

I got news of you through Bernie. No letters have arrived, so I assume that none of mine have gotten through either. I hope this one does better.

Did you ever get the pictures we had taken at Kingman? I think they may be pretty funny. The guy was so elaborate about it. Real professional pride. And, by the way, did Tommy L. [*Tommy Lauritsen*] deliver my blue suit?

You seem to be getting along fine with the pooch. I'm very glad you got him. Is he growing visibly?

One of the difficulties of life here is that it takes a couple of weeks to get admitted to the 7am Locker Club, which is the agency through which officers get liquor. I've been sponging, chiefly from Jim and Peer, but I hope to get some legitimately soon.

One or two items of island life: We do get our laundry done, and it only takes 3 days. The GI movies (combat films & news) are quite interesting. The movie is open air. It's standard practice to bring raincoat & helmet. The whole audience sits stolidly through the regular-as-clockwork evening downpour. When it rains it's like nothing you've ever seen—like being in a real strong shower-bath. I picked up a slight sunburn, not enough to hurt, but enough to do some peeling. [*For a small fee one could get laundry done by the Navy Seabees, who built ingenious windmill-driven washing machines. They also sold cowry shell necklaces.*]

Lots of love, Bob

Three days later came the Nagasaki drop. Phil Morrison, Luis Alvarez, and I wrote a letter addressed to Sagane, a Japanese physicist whom we had known in Berkeley. We taped the letter to the pressure cylinder that was to be dropped. A remarkable thing about the Japanese is that, after the letter was recovered and turned over to Japanese Naval Intelligence, since it was addressed to Sagane it was delivered to him. He eventually got it back to Luis Alvarez.

Headquarters
Atomic bomb command
August 9, 1945

To: Prof. R. Sagane

From: Three of your former scientific colleagues during your stay in the United States.

We are sending this as a personal message to urge that you use your influence as a reputable nuclear physicist, to convince the Japanese General Staff of the terrible consequences which will be suffered by your people if you continue in this war.

Fig. 5.8 Luis Alvarez with the letter to Professor Sagane written by he, Phil Morrison, and myself, after it was returned to Luis. (Photo courtesy of Gene Lester)

You have known for several years that an atomic bomb could be built if a nation were willing to pay the enormous cost of preparing the necessary material. Now that you have seen that we have constructed the production plants, there can be no doubt in your mind that all the output of these factories, working 24 yours a day, will be exploded on your homeland.

Within the space of three weeks, we have proof-fired one bomb in the American desert, exploded one in Hiroshima, and fired the third this morning.

We implore you to confirm these facts to your leaders, and to do your utmost to stop the destruction and waste of life which can only result in the total annihilation of all your cities if continued. As scientists, we deplore the use to which a beautiful discovery has been put, but we can assure you that unless Japan surrenders at once, this rain of atomic bombs will increase manyfold in fury.

On this mission I was supposed to be aboard the third plane, where my job was to run the Fastax camera. The Fastax on the Hiroshima mission hadn't worked, but some changes had been made in the switching, which Bernie Waldman had taught me. The Fastax wasn't complicated, but you had to know which switch to push when. The camera only ran for two or three seconds before using up all its film, so I had to know the time it took the bomb to drop—forty-three seconds at 30,000 feet—make a mental correction if the bombing plane happened to be higher or lower, start a stopwatch when the bomb dropped and start the camera one second before the explosion.

The drill was the same as for the first mission: crew briefing, breakfast at two in the morning, takeoff at three. We taxied to the runway, turned around, and the pilot revved up his engines. Then he called for a parachute check, and sure enough there was one parachute short. I don't much doubt that it was my parachute that was missing because I was the greenhorn, I didn't know the drill, and the supply sergeant probably had neglected to give me one when he gave me a lot of other junk. The pilot ordered me put off the plane. That was truly idiotic: he forgot that he wasn't on a joy ride, the plane was supposed to have a mission. The mission was to take pictures, and I was the only one aboard who knew how to run the camera. I tried to point this out to him, but those planes didn't have any soundproofing. The engines were making such a racket there was no possibility of making myself heard. The pilot and crew had throat mikes and earphones. So I started to wave my arms and a sergeant grabbed one, opened the door, and pushed me out. The plane took off, and there I was at the end of the runway at three o'clock in the morning, three miles from anyplace. Everybody knew there were Japanese kamikaze soldiers around who used to come out at night.

I walked back and finally got to our headquarters and walked in, much to the surprise of Gen. Thomas Farrell, who was Groves's man out there, and Colonel Tibbets and Luis Alvarez, who also happened to be there. Of course Tibbets threw a fit when he heard what had happened, and after a little conference with the general, decided to break radio silence to the plane. He gave the pilot a piece of his mind, then he put me on the phone to try and tell them how to run the camera.

That wasn't the only sin that pilot committed that night. He missed the rendezvous over the coast of Japan, and while the other two planes were down at Nagasaki he was up at Kokura, a couple of hundred miles away. He flew down to Nagasaki in time to see the mushroom cloud and the only picture we got was taken by his tail gunner with a snapshot camera.

Aug. 15 [*Tinian*]

Dear Baby:

Well, it's all over now. The news came in a couple of hours ago. It's nice to think that we, as well as the Russians, had something to do with ending the war. It was a good clean ending—it would have been terrible if it had dragged on for months of destruction.

There's surprisingly little excitement or jubilation here. The army seems to be taking the news quite soberly. I suppose one of the reasons is just that negotiations dragged on so long everybody got tired of it. There is no sign at all, so far, of any celebration.

Last night we watched the last strike returning. Headlights stretching to the horizon, like cars coming back to a city on a Sunday night. It goes on for hours and hours—it is incredible to see. Everyone had the same hope—that no one would ever see it again.

Some of the boys will be returning home in a few days. I'll try to send this letter with one of them, so maybe you'll get it. The only word I've had from you remains what Bernie brought. I'll be delayed a couple of weeks in returning. There's a rather unpleasant job still to do. (I am told the suggestion that I do it came from Opje.)

We had a big press conference here a few days ago. Did you hear me on the radio? Probably not; I think it was Columbia that I talked for. I wonder what is happening back at Y [*code name for Los Alamos*]. Is the place disintegrating? Is everyone trying to get home for the Fall semester? Opje must have his hands full right now.

We're expecting Henry to arrive this evening. It will be good to see him. I'm glad he got a chance to come, even if I am left without a raincoat.

We haven't been very busy for the last few days, and we've gotten a lot of swimming in. We have diving masks, and deep diving, in these clear waters, is a wonderful sport. Down along the coral cliffs, brilliant in the sunlight, with the breakers crashing into foam twenty feet overhead.

Luis and the other boys will no doubt give you a more detailed account of my inglorious adventures. [*The Red Cross supplied masks and snorkels. There was one unusual thing about the beaches at Tinian that I've never seen anywhere else; the sea floor was littered with 50-caliber machine gun bullets. It was really quite amazing how much ammunition had been fired in the course of the invasion.*]

Lots of love, Bob

Peace! Let me tell you, we were really heroes out there in the Pacific. There were an awful lot of guys who weren't looking forward to landing on the Japanese beaches in October. And there were about three million men whose main desire in life was to get back home. They thought we were great.

[*Tinian*] Sunday, Aug. 27 [*Error: Should be August 26.*]

Dearest Charlotte:

The last letter I wrote you is still reposing in Luis's briefcase, but even so it seems the more likely way of getting mail home. The boys are all very anxious to leave, and they are having a tough time of it. They were supposed to go last week but the Commodore (ex-Captain) [*Captain Parsons, who had been promoted*] got jumped on by Nimitz, Spaatz, and LeMay separately—saying that they have to stay [*Adm. Chester Nimitz and Gen. Carl Spaatz; Spaatz was the general commanding all strategic bombing in the Pacific*]. Now they are mad because a telecon just arrived from Y, signed by Kisty [*George Kistiakowsky*] & [*Norris*] Bradbury saying to stay & what to do about the next one. Which strikes everyone as being rather silly.

Henry was here for about 24 hrs. You (and more particularly Shirley) should have seen him. He is now a big wheel, as they say in the air forces. Ordering people around, looking harried and important, etc. Henry & Warren [*Stafford Warren, a radiologist and colonel in the Army Medical Corps*] are going to Nagasaki, with one group [*a group of SEDs—Special Engineering Detachment—sent from Los Alamos; equipped with Geiger counters, their job was to measure radioactivity on the ground at Hiroshima and Nagasaki*]. I have a date to meet Henry there. He brought lots of news—how things are being taken at Y in particular. No civilian here has received any mail, so Henry was a godsend. We saw him off one evening—embarking on a ship which was to join up with a Marine division—to land at Nagasaki. Next day we discovered there had been a slight mixup, and they'd left on the wrong ship—the one they were on was headed for Tokyo. What finally happened was that they had to be taken off in mid-ocean by a destroyer. What the

Fig. 5.9 New York Times reporter William Lawrence, Henry Barnett, and I at Tinian.

destroyer did with Henry is still unknown. Meanwhile a letter arrived for him from Shirley. I've taken it and will deliver it personally—its the only chance he has of getting it.

I was, and still am, scheduled to leave for Tokyo on Tuesday. However the radio this morning says that MacArthur has delayed his landing till Thursday, so we will probably be held up. Maybe we'll fly to Iwo Jima first and wait there. We have work to do in Tokyo, Hiroshima, Nagasaki, and assorted other places. It would take 3 weeks to do if there is no holdup, which is not to be expected. So it seems unlikely we'll be back before October. Phil is in charge of radioactive surveys, & Bill Penney in charge of damage studies. There was nothing obvious for me to do, so I was made head of the mission.

Life here has reached the beachcombing level. Nothing to do but eat, sleep, and swim. We all survive pretty well on such an existence.

There's a prize story about Spaatz's last visit. Baker was showing him some of the works. Spaatz questioned something Baker told him, Baker repeated his explanation, Spaatz said, "You may believe that, but I know what I believe," turned on his heel, and marched out in a military manner. [*A little later I was talking to one of Spaatz's staff officers and he said, "You know, the General is very unhappy about this development." I was quite surprised, I thought it would be the opposite. "No." he said "the General says, 'Now we'll only need two or three planes.' "*]

The latest fad here is trading with the natives, soap for cowry shells.

This will undoubtably be a continued letter. I'll keep working on it till we finally leave.

Thursday: Since starting this letter one item has been discovered. The reason we haven't been getting mail is that it's been going to the 1st Tech. Supply Det. in Manila. Also, since the war ended, no air mail is going east. V mail is reputed to be moving so I wrote one a couple of days ago. I'll give this to Charlie Baker, who is to leave tomorrow.

The tornado [*I meant typhoon*] that hit Japan and held MacArthur up manifested itself here in torrential rains. We've been wading in mud for three days, and it looks as if we're due for some more. Which reminds me of another item of local life—we haven't felt hot water since we left Oahu. But it's really not missed in a tropical climate. The bathing facilities consist of the Pacific Ocean, plus very nice outdoor showers.

Believe it or not, we still go to the movies every night. The pictures are not too good, being about 90% musicals. However it's important to have some kind of routine to fill in the day. One of the other amusements is listening to the radio. We have very good short wave sets, and get anything between San Francisco and Delhi. The Tokyo news during the critical days has been interesting to listen to. The tone is conciliatory, and democratic reforms are being promised—these are the main lines. There's a pretty good library here. I'm reading Sandburg's Life of Lincoln.

Phil is leaving for Okinawa in a day or two. I haven't any further news on when I'll be moving.

With much love, Bob

August 31 [*Tinian*]

Dearest:

It now seems that V mail is the only way of getting letters east. I'm sure you haven't been getting any more than we do here, which is nothing.

We've been having a quiet time—just waiting. We were due to leave for Tokyo today, but it's been postponed. There's no telling when we'll actually get there. Meanwhile it's pleasant enough: sleeping, eating, swimming.

CHAPTER 1   THE MANHATTAN PROJECT

THE LOS ALAMOS TEAM AT TINIAN ISLAND FOR THE HIROSHIMA AND NAGASAKI
NUCLEAR WEAPON STRIKES IN AUGUST 1945

PHOTOGRAPH AND NAMES COURTESY OF HAROLD AGNEW

LEFT TO RIGHT:
FRONT ROW: (– –), ART MACHEN, ROGER WARNER, HARLOW RUSS, NORMAN RAMSEY, ED DOLL, ENSIGN JOHN
              TUCKER, ENSIGN GEORGE REYNOLDS (?), MILO BOLSTAD

SECOND ROW: CHARLES BAKER, PHIL MORRISON, BILL PENNEY (U.K.), TED PERLMAN, TOM OLMSTEAD,
              COMMANDER A. F. BIRCH, ADMIRAL PARNEL, GENERAL T. F. FARRELL, CAPTAIN W. S. PARSONS,
              COMMANDER RICHARD ASHWORTH, BOB SERBER, LARRY LANGER, BERNARD WALDMAN,
              LUIS ALVAREZ

THIRD ROW: (– –), (– –), HENRY LINSCHITZ, (– –), RAYMER SCHREIBER, WALTER GOODMAN, (– –), (– –), (– –),
              HAROLD AGNEW, LARRY JOHNSTON, LIEUTENANT COMMANDER E. STEVENSEN,
              ENSIGN D. L. ANDERSON, ENSIGN DON MASTICK (PARTIAL)

BACK ROW: MORT CAMAC, TECHNICAL SERGEANT KEN KUPFERBERG, ENSIGN BERNARD O'KEEFE, (– –), (– –), ...

NOT PICTURED: SHELDON DIKE, CAPTAIN J. F. NOLAN, M.D.

1–34a

Fig. 5.10 The Tinian team. (Photo courtesy of Harold Agnew)

Henry was here for a day. Very important, in charge of his detachment. He left by ship, but somehow got put on the wrong boat and had to be taken off by a destroyer in mid-ocean.

Phil brought a copy of *Time* back from Guam the other day. The boys were all delighted to see radar pushed into the back pages. [*The story of how scientists had played a heroic role in the war effort thanks to the development of radar had been all over the newspapers. Nothing, of course, had been printed about us.*]

It seems unlikely that I'll get back before October.

Lots of love, Bob

Sunday, Sept. 2 [*Tinian*]

Dearest:

Well, it's all officially over. We heard the signing and Truman's radio speech this morning. I heard it from a very appropriate point: Cincpac headquarters in Guam. Cincpac means Commander in Chief of the Pacific—it's Nimitz's hangout, and his five star flag was flying right outside our conference room. To start a little earlier in the morning, I went down from Tinian to Guam with the Commodore for a conference with a couple of Admirals. Luis & Bernie and I had been planning to go to Guam anyway today, just to see what it's like. We intended to ride TAG (Tinian & Guam), which is what the air transport line is called. So it was very convenient when the Commodore dreamed up this conference, and provided one of our C-54's, one of the "Green Hornets," to take us down. Luis and Bernie came along for the ride. We left at 9:30. It's about 125 miles to Guam, a nice 40 minute trip.

It's a beautiful ride and a beautiful sight coming in over the ocean to a very green tropical island, with the water shading from a light turquoise on the beaches to a brilliant blue offshore. The palm-trees are lovely seen from the air; a grove of trees looks like a bank of little green flowers—on the ferny side.

Guam looked like a tropical isle is supposed to look—palms, flowers, the most luxuriant kind of vegetation. Of course one had to ignore the occasional shell shocked houses, fortifications, etc. And the army & navy installations which look the same everywhere. However we saw the best of it. Nimitz put his headquarters on top of the highest hill on the islands, and his view is magnificent. Blue water, green palms, and a cool breeze.

We had the conference during the signing of the surrender, which we could dimly hear on the radio in the next room. Then we quit in time to hear Truman. Then lunch in Navy style—which sure puts the Army to shame. This was the mess for senior officers—tablecloths, waiters, clean silver—very swank and quite good. Cold meats, potato chips, pickles, iced cocoa, fruit salad. And after lunch back to Tinian.

Now that the war is over I don't see why it isn't ethical to back track and tell you a little more about our trip out here . . . [*The ensuing has been moved to the beginning of this chapter.*]

We're still sweating it out. No news of any movement, east or west.

With love, Bob

SIX

Hiroshima and Nagasaki, 1945

Yokohama, Sept. 9

Darling:

We seem to be having a hell of a time with our mail. The first letters I got from you arrived just a few hours before we left Tinian. By this time you will have gotten from me at least the letters sent by courier. It was wonderful to hear from you after so long. Two letters, one dated July 27, and one Aug 25. Your remarks about the exciting things going on at home seemed quite strange out here. We are completely insulated from such feelings. We have no feeling of either influence or importance. The possibility of a job at Berkeley is very attractive. I know how much you want to get away from Urbana, and that is a very strong argument. And of course the Cal dept. isn't a bad place to be either.

We left Tinian at midnight Friday night, gassed at Iwo Jima, and came on to the Empire. We didn't see much of it—it was covered by a fog bank. We had to circle above the fog, the airport, for a couple of hours waiting our turn to land. The traffic (troops in, PW's out) is enormous. The Japanese countryside, seen by air and jeep, is very lovely. Each field, each growing thing, seems to have had the most anxious care. The woods look like a carefully kept park. The varieties of trees seem much prettier than ours (maybe I'm prejudiced by 2 $\frac{1}{2}$ years of New Mexico). We have heard so much of the Westernization of Japan that I was quite surprised to see how little evidence there was of any such thing in the countryside, in the tools, clothing, housing, or (I imagine) way of life. Oxen drawn carts, picturesque straw roofed houses, pajamas & kimonos. Close up the houses look dirty and poverty stricken and incredibly flimsy and transparent. The kids are cute and friendly, trying to sell eggs and tomatoes. Many of the men are in fragments of uniform. The people seem reserved but not unfriendly. Everyone says the Japs do everything they can to be helpful. It also seems remarkable that a people who have just a two lane dirt road from Yokohama to its principal airport thought they could win a war with us. The city is hell. Huge areas completely gutted, the people existing in shacks thrown up of reclaimed sheet iron. Red rust is the color of everything. Parts of course are untouched. We sleep next door to the Grand Hotel, and eat at the Grand. The food isn't anything special, but the service is superb—tablecloths, Jap waiters, just like it must have been in the tourist days. There is no visible economic life— no factories running, no stores. What the people find to live on is a mystery.

Tomorrow we hope to move our quarters to Tokyo. It looks very good from the point of view of getting our work over quickly. I hope we'll be finished in 10 days, and home in two weeks. Let me say again how fine it was to hear from you, and how little I understand what is going on back there.

All my love, Bob

Of course I was unduly optimistic. We were on our own and didn't know how to arrange transportation to Hiroshima and Nagasaki. Somewhere we heard that Adm. Richard Byrd, the polar explorer, was in town on a roving mission for the navy, with his own plane. Someone suggested that the admiral might be persuaded that he wanted to see Hiroshima and Nagasaki. That worked—though we paid a price for it, for Byrd took a liking to Ensign Reynolds (George Reynolds, a young physicist in our group, had applied for a commission in the Navy, which came through just as peace was declared) and decided to keep him as an aide. We got a pass to Yokohama Airfield which identified us as "Admiral Byrd's Party."

Yokohama, Sept. 11, (5 pm)

Pussy:

Our business is not advancing very rapidly. This morning we left by plane—we were going to pick up Warren and Jim [*Stafford Warren and Jim Nolan*] at Hiroshima, fly to Nagasaki, where Henry [*Barnett*] is due to arrive tomorrow. We couldn't land at Hiroshima because of fog, flew on towards Nagasaki, finally got lost, had to circle down through the fog to within a couple of hundred of feet of the coast so the pilot

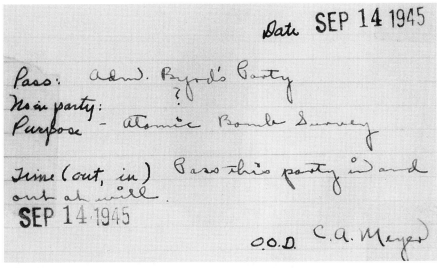

Fig. 6.1 Our "pass" into Nagasaki.

could find out where we were, turned around and came back. A nice little jaunt of 1,300 miles. We'll try again tomorrow. And guess who we had for a passenger: Admiral Byrd. A handsome, unintelligent looking guy. The opinion of him seems to be that he's one of the best families and does no harm.

The most striking impression continues to be the complete breakdown, bankruptcy, destitution of everything in Japan. The stores are empty. Our people go around with their tongues hanging out trying to spend money, but there just isn't anything left. The few Japanese cars and buses have a huge contraption on the back to burn charcoal, wood, garbage, on which the car manages to run.

I was handed a letter from you this morning on the plane. It was the first one you wrote after I left. How is Sawdust? Sliderule? [*Charlotte's horse and dog*] your ankle? Have you been riding? These are rhetorical questions, I hope to be back before you have time to answer.

Lots of love, Bob

Nagasaki, Sept. 20

Dearest:

It's a week since I last wrote from Yokohama. As usual, we've daily been expecting to leave here, and I thought a V mail letter from Tokyo would be a lot quicker than air mail from Nagasaki. But since it has

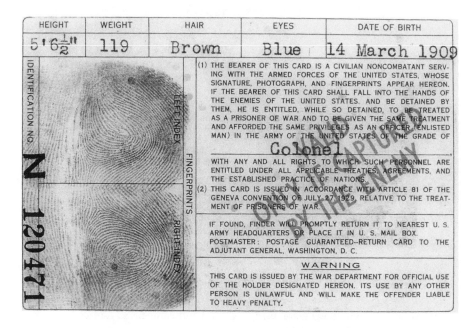

HEIGHT	WEIGHT	HAIR	EYES	DATE OF BIRTH
5'6½"	119	Brown	Blue	14 March 1909

IDENTIFICATION NO. N 120471

LEFT INDEX

FINGERPRINTS

RIGHT INDEX

(1) THE BEARER OF THIS CARD IS A CIVILIAN NONCOMBATANT SERV-ING WITH THE ARMED FORCES OF THE UNITED STATES, WHOSE SIGNATURE, PHOTOGRAPH, AND FINGERPRINTS APPEAR HEREON. IF THE BEARER OF THIS CARD SHALL FALL INTO THE HANDS OF THE ENEMIES OF THE UNITED STATES, AND BE DETAINED BY THEM, HE IS ENTITLED, WHILE SO DETAINED, TO BE TREATED AS A PRISONER OF WAR AND TO BE GIVEN THE SAME TREATMENT AND AFFORDED THE SAME PRIVILEGES AS AN OFFICER (ENLISTED MAN) IN THE ARMY OF THE UNITED STATES OF THE GRADE OF

Colonel

WITH ANY AND ALL RIGHTS TO WHICH SUCH PERSONNEL ARE ENTITLED UNDER ALL APPLICABLE TREATIES, AGREEMENTS, AND THE ESTABLISHED PRACTICE OF NATIONS.
(2) THIS CARD IS ISSUED IN ACCORDANCE WITH ARTICLE 81 OF THE GENEVA CONVENTION OF JULY 27, 1929, RELATIVE TO THE TREATMENT OF PRISONERS OF WAR.

IF FOUND, FINDER WILL PROMPTLY RETURN IT TO NEAREST U. S. ARMY HEADQUARTERS OR PLACE IT IN U. S. MAIL BOX. POSTMASTER: POSTAGE GUARANTEED--RETURN CARD TO THE ADJUTANT GENERAL, WASHINGTON, D. C.

WARNING

THIS CARD IS ISSUED BY THE WAR DEPARTMENT FOR OFFICIAL USE OF THE HOLDER DESIGNATED HEREON. ITS USE BY ANY OTHER PERSON IS UNLAWFUL AND WILL MAKE THE OFFENDER LIABLE TO HEAVY PENALTY.

Fig. 6.2   ID issued by the War Department just before I left for Tinian.

finally become clear that I'll be here for three more days, I'll try this way. When I wrote last we'd just gotten back to Yokohama from an unsuccessful attempt to get here. The next day the weather was too bad for flying, so we moved to Tokyo and got rooms in the Dai Ichi Hotel (next best to the Imperial, a big modern joint). [*We went to MacArthur's headquarters and were almost scared to death at the entrance by the elite guard that snapped to attention and presented arms as we approached the door. Inside, we were asked for our IDs. Mine said, "Name: Robert Serber" and below, "Designation: Colonel"; diagonally across it was written in large red letters, "Valid only if captured by the enemy" (see figure 6.2). But MacArthur's headquarters was big on protocol and there the designation "Colonel" was valid. So I was given a room in the Dai Ichi; generals went to Frank Lloyd Wright's Imperial Hotel.*] The trip from Yokohama to Tokyo runs thru miles and miles of burnt-out worker's houses. The factories are relatively untouched, but all houses are gone—the people live in makeshift shacks built out of rescued sheet iron. The center of Tokyo is a god awful mess. Along the Ginsha (5th Ave) half the buildings are heaps of rubble. When you look, you see the other half are gutted by fire. The few stores still open have nothing in them I was told. Saw Phil in Tokyo.

Fig. 6.3 In Tokyo, with a water pump—what the Japanese had left as a fire engine.

The next morning we tried for Nagasaki again. Flew at a couple of hundred feet along the coast so we wouldn't get lost again. It worked, and made a very scenic trip. The coast is beautiful: green rugged irregular shore line, lots of little islands offshore. One propeller went flooie on the way, and we blew two tires trying to land a C-54 on a 1,000' runway, complicated by three parked Jap planes. We overshot quite a bit, and bounced along on the grass. Then it was 20 miles to town in a charcoal burning bus that made about 3 mi/hr, and had to be helped over the mountains. We are staying in the Beach Hotel, which in pre-war days catered to foreigners. It's a very pretty Japanese cottagy sort of place in a little fishing village, Mozi, about 8 miles from Nagasaki and right on the sea. The incredible part of it is that most of the time Penney and I have been out there alone, 8 miles beyond the nearest American outpost, and completely unarmed. But nobody payed any attention. In fact, after a couple of days even the Japanese police guards got bored and went home. We were invited to lunch a couple of times by the captain of the U.S.S. *Haven*, a hospital ship lying at the dock, the captain being prompted by Admiral Byrd. It's pitiful as hell to see the prisoners come in and hear their stories—mostly of callousness, starvation,

Fig. 6.4 Harry Whipple (a medical officer), myself, and Henry Barnett at Nagasaki. (Photo courtesy of Henry Barnett)

and slave labour. After a few trainloads of PW's any latent sympathy for the Japanese is pretty well dispelled. The people here are sullen, obliging, and one feels not too friendly. I'll skip any description of our manufactured hell.

We've wondered all week what the hell had happened to Henry, and finally found a Marine colonel who had, with great difficulty, finally located him on Okinawa, where, literally, he'd been forgotten for a month. But finally, last night, Henry, Warren and party showed up. I delivered 4 letters from Shirley which I'd brought from Tinian. Jim Nolan and Whipple [*Harry Whipple*] arrived too. Henry is in fine spirits now that he's here and has work to do. He spent the month growing a moustache.

My present plan is to leave here in about 3 days for Hiroshima, get a couple of days there, and start home the 26th. So I'll probably beat this letter in.

Lots of love, Bob

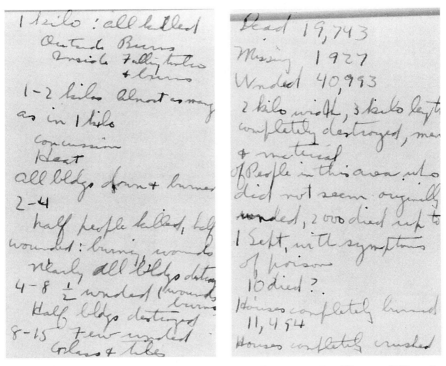

Figs. 6.5 and 6.6   Notes of my interrogation of Japanese Intelligence Officer, in Nagasaki.

[*P.S.*] I forgot to mention the typhoon, which held us here a couple of days extra by cutting us off completely.

Our first encounter with the prisoners of war came one day Bill Penney and I were standing next to the railroad tracks alongside the Munitions Works when a passenger train came through. It was full of POWs who had just been released by the Japanese, and they hung out the windows yelling and waving, great excitement, and we realized that we were the first free white men they'd seen in years. They were mostly Australian and Dutch taken in Singapore and Indonesia early in the war. They were the survivors, living skeletons.<6.5–6.6>

On September 13, our first day in Nagasaki, I interviewed a Japanese Naval Intelligence officer, a Lieutenant Likura. We sat on a flat rock overlooking the ruins while I copied in a small notebook as he recited it the official inventory of damage. (See figures 6.5 and 6.6.) The next morning Bill and I started to examine the damage for ourselves.

Fig. 6.7 Nagasaki.
Fig. 6.8 Masonry building.

The first impression: everything flattened and burnt over in the residential and small business area. A macabre thing, there was only one item that survived the blast and fire, and that was office safes, those heavy iron safes about two feet square and three feet high. They stuck up above the ruins and the surprising thing was how many of them there were.

Bill was quite ingenious designing ways to determine the shock wave pressures. At various distances he found five gallon gas cans that had been more or less crushed by the explosion and he took them back to England where they made similar cans and measured what pressure it took to squash them similarly. Then he did things like take concrete from the remains of reinforced concrete buildings, which could be tested for strength. One day he decided to see what the distant damage was like. He got a jeep and an interpreter and they drove off several miles until finally he found what looked like a perfect example, a door with paper panels, half broken, half intact. Through the interpreter he got hold of the woman of the house and they had a little conversation. Bill, pointing: "Atomic bomb?" Woman, "No. Small boy." That sounds like a made-up story, but it was really true.

I had been given a good camera—a Graflex, the kind that newspaper reporters carried in those days—and I took a lot of pictures. The walls of masonry buildings simply crumbled. The quarter-mile-long steel-frame building of the Munitions Works had completely collapsed on the machine tools within. Reinforced concrete buildings stood up best. I also studied the flash burn caused by the bomb's fireball. The heat radiated by the fireball was, for a second, at a mile distance, a thousand times that of the sun. The sides of all the telephone poles facing the explosion were charred. One could follow the line of charred poles out beyond two miles from ground zero. At one point I saw a horse grazing. On one side all its hair was burnt off, the other side was perfectly normal. Later on, when I was back in Los Alamos, I was giving a seminar on what I had seen in Japan and I remarked that the horse was happily grazing, and Oppie scolded me for giving the impression that the bomb was a benevolent weapon. I also found, just about a mile from ground zero, a wooden crate, a very light kind of crate that I'd call an orange crate, with thin wooden slats that had been set on fire by the flash. The fire got a good start, it was curling around the back sides of the

Fig. 6.9 Nagasaki Munitions Works, a light steel-frame factory building.
Fig. 6.10 Reinforced concrete building, the kind that stood up best.

Fig. 6.11 Church wall still standing in Nagasaki.

slats and through the nail holes when evidently it had been blown out, presumably by the blast which hit five seconds after the flash. At one mile the wind velocity behind the blast front would be about 170 miles an hour. I lugged that crate all the way back across the Pacific to Los Alamos and I wrote a report about it. I turned it over to the Photography Department to have pictures taken. The next morning a very sheepish photographer came back, very apologetic. He said he had set it up in front of the camera with all the lights and what not and just then five o'clock came and he had left it to finish the job the next morning. Of course, during the night the janitor saw this piece of trash and threw it out.

The city of Nagasaki was divided in two by a row of hills. The sector containing the heavy industrial plants was completely destroyed. The other sector was partly shielded by the intervening hills. As mentioned in the following letter, some stores, a bank, and the red light district survived.

Omura, Sept. 27

Dearest:

Letter writing at the moment is discouraged by the fact that we have no facilities for getting them mailed. However, I'll write an account of some of our Nagasaki doings, and save it till I find a mailbox.

As I wrote before, we came down in style with 2 generals, 1 admiral, and about 5 useless colonels. After they left Bill Penney and I stayed on alone at the Beach Hotel, 8 miles from Nagasaki. We had a fine time, got quite a lot of work done. The Japs are extremely anxious to be helpful and do anything you ask of them. The people are curious, often frightened, sometimes indifferent, but never visibly unfriendly. It is an attitude hard to understand; neither side would behave anything like the same way if the positions were reversed. They—all the unimportant people—were firmly convinced they were winning the war up to the very moment of the Emperor's surrender.

As illustrations of the helpfulness of the Jap officials: our transportation, in the beginning, was an exceedingly broken down charcoal burning bus. We asked if a passenger car were available. None were visible anywhere. By the next morning, they'd produced a Dodge sedan for us. The following day a big Buick appeared. And the third day and thereafter we rode in real style in a huge Cadillac limousine about half a block long. Another example was at the torpedo plant, where we blithely requested the chief engineer to chop a hole in the concrete floor of one of his factory buildings. He promised to do it if he could find a hammer and chisel (illustrating how much the Japs have left), and sure enough, when we came back two hours later there was the hole in the floor.

Henry, Warren, and company arrived about the end of the first week. Henry and his gang had had a terrible time, completely forgotten by everyone on Okinawa for a month. They immediately got to work at the local hospitals, were quite discouraged at first by lack of records on the patients etc, but have dug up more and more, particularly at the Naval Hospital at Omura, and are really enjoying the work. Henry and I went out on a survey one day [*a survey of the city for radioactivity, by Henry's Special Engineering Detachment with their Geiger counters*], and that end is also going quite well, after a slow start.

One night we were all invited for dinner on the cruiser *Wichita*, and it turned out to be quite an event. I never saw a bunch of people go to so much trouble to see that the guests had anything they wanted. Turkey and pie à la mode for dinner, then they opened the ship's store for us, opened the soda fountain, gave us hot showers, offered to do our laundry, flattered us by listening to atomic bomb stories all evening. Warren gave a talk, and they showed movies, and finally we stayed overnight. The Navy has it all over the Army—a ship is (comparatively) just like home.

We were lucky in one way—we had an unusual opportunity to see Japan uncontaminated by occupation troops. We arrived first, the next day a small fleet appeared, escorting a couple of hospital ships to take off POW's, but the naval personnel weren't allowed ashore at all. Their marines held the dock and that was as far as they went. [*They were awaiting the completion of our survey of radioactivity and our certification that it was safe to land. This was held up by Henry's late arrival.*] After about ten days the stores began to open up and we scrounged around picking up souvenirs. This was made difficult by the fact that we didn't have any yen, and had to persuade them to take dollars. We finally got across the conception that $1=Y15 but it unfortunately turned out that the Japanese bank would only pay the pre war 4.25 Y = $1 rate. The most nearly troubled moments I have had in Japan were in being followed by a group of irate shopkeepers waving dollar bills they thought they'd been gypped on. Henry and I made a fine shopping team: I would explain the mathematics of the exchange ratio (an illustrated lecture, with props) while Henry encouraged them to believe in our honesty by prominently displaying the butt of his .45.

But then, a couple of days ago the 2nd Marine Division moved in. A tough lot, who promptly declared all the stores out of bounds. Our first real indication of the changed times came, however, the night before last. Our party had been invited to dinner by the Governor of Nagasaki Prefecture at the best geisha house. We had a fine time, squatting barefoot on little mats, with a very excellent dinner (which I managed quite well with chopsticks). Roast beef, cut in narrow strips, sukiyaki (meat & vegetables), meat balls and vegetables. The geisha girls did some graceful and colorful dances, with masks as the principal props. Then they were kneeling on the floor behind us pouring saki into our little porcelain cups—a very pretty Japanese scene—just at the moment when the MP's broke in. About half a dozen 6 foot Marines covered the doors. Their lieutenant said, "Gentlemen, you are all under arrest. May I speak to the senior officer present?" And the thing that made it so ludicrous was the complete formality and grimness, from battle helmet to .45's at their belts to—stockinged feet. Even the MP's had to remove their shoes before entering, and it put them in an unhappy position which they didn't enjoy. It got even worse when they headed us to the doors, since all GI shoes look alike. And the MP's

Fig. 6.12   At a party given by the governor of Nagasaki, before interruption by barefoot American MP marines.

had to take whatever was left over. The Governor was very unhappy about it, and Col. Warren got bawled out by the Commanding General the next morning for fraternizing with the natives. Poor Warren even missed the dinner, he arrived just after the joint was raided.

Now to try to explain what I'm doing in Omura, which is the town, about 40 miles from Nagasaki, where the airfield is located. General [*James B.*] Neuman left about a week ago, taking Penney & Reynolds with him, and with the understanding he would come back in three days, pick up Col. Warren and me, and take us to Hiroshima. We haven't heard a single word from him since then. Last night I was telling this to an aircrew that ran out of gas, had to stay over at Omura, and somehow had gotten to the Beach Hotel for the night. After talking it over we decided that what I needed was a personal C-47 and crew (they were anxious not to have to return home to Okinawa). They looked at my orders, which say any theatre, any transportation, and are signed by the Office of the Secretary of War. They said that covered anything, so, since (somewhat to my surprise), Col. Warren acquiesced, we came up to Omura today and radioed Okinawa asking that the plane be assigned to our mission for a few days. No answer has come back yet, but if the gag works, I'll fly up to Tokyo tomorrow (weather permitting) to find out what the score is and get a clean shirt. Meanwhile it's very convenient, since Henry is staying with his group at the Naval Hospital in Omura. It's a huge place, comparable to Bruns [*the V.A. hospital in Santa Fe*], in command of a cute little Japanese admiral, who is as friendly as can be, thinks Henry is wonderful, and supplies excellent food and quarters, including real silk mosquito nets. I haven't looked at the sheets yet, maybe they're silk too. So tonight, for the second time since leaving the States, I'll get a hot bath, this time Japanese style. We brought the two air crew lieutenants over for lunch. They were overwhelmed by a glass of milk each. The first time they'd seen the stuff for 18 months.

I guess that hits the high points here. I haven't had any mail of course since leaving Tokyo, there being no mechanism for getting it. Meanwhile, lots of love, and it still looks like I might get back in the first half of October.

Love, Bob

I omitted a description of our trip to Omura, probably so as not to alarm Charlotte. The plane that had brought us to the fighter strip near Nagasaki had finally gotten new tires to replace those it had burst in landing, and the pilot offered to fly us to Omura. So Colonel Warren, a couple of his henchmen, and I drove to the strip in the colonel's jeep and loaded everything aboard the C-54. But at this point the pilot got cold feet and decided he couldn't take off with all that weight aboard. So jeep, Warren, and henchmen were unloaded to make their own way to Omura. I stayed aboard for the hell of it. We made a carrier-type takeoff—brakes locked, engines revved up full, brakes suddenly released—and we went roaring down the runway. We lifted off at the very end.

The survey found no radioactivity at all in the city. One night Henry and I were plotting out on a map the day's readings and we noticed a high spot. The boys went back there the next day, searched around, and found a radon needle that had evidently come from a cancer patient.

The ruins were hard enough to endure, but the really harrowing experience was a visit to a Nagasaki hospital with Henry. It was a makeshift hospital, a building with the front wall blown out, the patients on cots inside and on stretchers outside on the ground. This was five weeks after the bombing and the patients were mostly suffering from flash burn or radiation sickness.

Tokyo, Oct. 1

Darling:

I'm sorry to have to report that our job won't be done for another week. I know it must be very hard for you back there to understand why it takes so long and is so hard to get anything done. The difficulties and snafu have to be seen to be believed. If Bill and I were here alone, armed with a couple of authoritative letters, we could proceed much more quickly and efficiently than we can weighted down with a whole military mission of 40 or 50 men to get in the way and ball things up. Here's the history of how things have gone since Farrell left us in Nagasaki. Neuman was to come back in two days, but didn't show up for a week. He took Bill back with him and was to return for me and Col. Warren in 3 days. Ten days went by and we didn't hear a single word from Tokyo. Not that both ends weren't sending messages—communications just broke down completely. I'd be there yet if I hadn't taken the initiative in my own hands and snagged a personal C47 to take me back to Tokyo, as I told you in my last letter that I was trying to do. When we got here I found that Nolan had gone down to Nagasaki the preceding day to get Warren and bring him back to Tokyo. They were expected the same day that we arrived, but didn't show, and in fact now, two days later, there's still no word of them. That's how things are here—as soon as anybody gets out of sight they effectively vanish from the face of the earth.

Meanwhile Gen. Neuman is sitting around waiting for Warren to get here before deciding on what to do about getting us to Hiroshima. Everything was at a complete standstill. I wrote a dispatch to Oppie yesterday morning and it took 32 hours to persuade Neuman to send it. He was waiting for Warren. This morning Neuman dismissed my C-47 pilot, but he hung around with us all afternoon. Bill and I finally decided the situation was too damn ridiculous: we had a plane, there was nothing to prevent us flying to Hiroshima and forging for ourselves. It seemed to us that all the objections and difficulties (food, water, transportation) that looked so big from an office in Tokyo, would turn out to be simple enough on the spot. We went in to see the general ready for a fight—picking up de Silva and [Hymer] Friedel on the way (of course not in a fighting position). I don't know what happened to Neuman. He seemed completely deflated, discouraged, and depressed. He acquiesced in the plan for keeping our plane and flying down tomorrow. We'll take Friedel and [Paul] Hageman along too. It should take 4 or 5 days. Then we shall be ready to come home.

To return to the plane trip from Omura to Tachakawa (Nagasaki airfield to Tokyo airfield). I had a wonderful time—copiloted the plane. The casual way we navigated was astonishing. The pilot never made a calculation or measurement, just judged the course by eye on his map, and checked whenever a break in the clouds showed something recognizable on the ground. After a while we got out of the clouds and had a lovely ride. The coast of Japan is the most beautiful I've ever seen. Mountainous, irregular, cut into hundreds of islands, promontories, isthmuses, capes. Every hill is wooded and terraced for agriculture, every little island is inhabited. Fuji is a fine mountain, sticking up through the clouds.

There was the other side too. We flew over Nagasaki, Hiroshima, Osaka, Kobe, Tokyo. It is hard to convey the utter ruin wrought by our B29's. The big cities are 70% or more destroyed. And every little town all up and down the coast has had hell knocked out of it. We circled Hiroshima for half an hour. It is far more impressive, in its way, than Nagasaki, for it is a dead city: a few concrete buildings standing, elsewhere only the paving of streets to show the rubble had once had some meaning. Not a living moving thing in all the miles of wasteland.

Yesterday's *Nippon Times* carried a story on the statement of the Oak Ridge scientists' association. Not a bad statement. Just about the same gospel we'd been spreading here when opportunity offered. [*The Nippon Times of September 30 carried an article mentioning the formation of the Association of Oak Ridge Scientists at Clinton Laboratories, and quoted at length from a statement it had issued calling for international control of nuclear weaponry.*]

Today I received letters from you dated Aug. 4th and 8th, giving some description of D + 1 [*the day after*]. It must have been quite a time, much more exciting (though not as tense) as ours.

Darling, I hope this is the last letter I don't beat home. I've thought that of a number of them before this, but our job now seems on a businesslike footing, and it shouldn't take long. As long as we, and not the army, keep the initiative. With much love, and high hopes of seeing you again soon,

Bob

There was no Beach Hotel near Hiroshima. Somewhere nearby we were given a large common room in a public building and tried to sleep on hard mats and wooden pillows. Our work in Hiroshima was a repeat of Nagasaki, with an interesting addition. In the Post Office Building, just a mile from ground zero, in a room facing the explosion the glass had been blown out of a large window, the beaver-

Fig. 6.13 Map of
Hiroshima and effects
of blast.

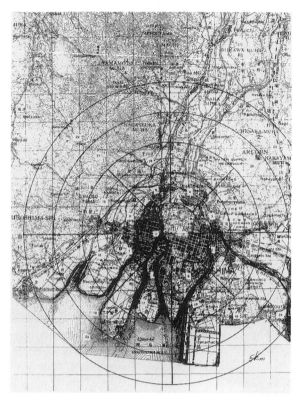

board of the adjacent wall all scarred by fragments of flying glass, but
the frames of the windowpanes were intact and their shadows were
clearly visible on the wall. The shadows were in reverse, where the
light hit it the wall was burnt black by the bomb's flash, so the shadows appeared white. By measuring the angle of a shadow, I could triangulate back and find out how high the bomb was when it went off,
which turned out to be 1,900 feet. By measuring the penumbra of the
shadow, I could get a rough idea of how big the fireball was.

While at Hiroshima we lost our plane and pilot. As you can guess
by now, when we finished our work Bill and I found ourselves
stranded. But also by now we were experienced in fending for ourselves. We found we could get a train to Tokyo. It was an overnight
trip on a stormy night. The train's rest room compartments had
Japanese-style plumbing. One was supposed to squat over a hole in
the floor open to the tracks. I found this difficult in a swaying coach,

Fig. 6.14 Hiroshima Post Office building, about a mile from ground zero.

so when the train made a scheduled stop at some intermediate place I crossed to the station in search of a more stable rest room. I found two doors, one marked with the image of a man, the other of a woman. But when I entered the one marked men I found that both doors opened into the same room. I took this as a case of attempted Westernization that had gone astray. As I was hurrying back to the train across the windswept platform, a man stopped me. I was surprised to see he was a European. He told me he was a Russian and had spent the entire war in Japan. He said he had a daughter at UCLA and begged me to try and let her know he was still alive. He gave me a letter addressed to her, in Russian with an English translation. When I got home I did succeed in speaking to her on the phone: I remember reading to her the English version of her father's letter.

Back in Tokyo we visited Yoshio Nishina at his laboratory at the university. He told us about the hard time he and his students had had during the war, and how they had had to scrounge around for everything they needed. To keep from starving they had to grow

Fig. 6.15   Room in interior of Hiroshima Post Office building, where the blast blew out the glass but left the windowframes intact, creating a reverse shadow effect. The wall was burned black by the light from the blast, while remaining white in the shadow of the windowframes. This provided fundamental information about the bomb blast. I could triangulate to find how high the bomb was when it went off (1,900 feet), and by measuring the penumbra of the shadow got a rough idea of the size of the fireball.

their own food on the university campus, and he proudly showed us their vegetable garden in the yard of the lab building. He and some of his students had been down to Hiroshima about three weeks before our own trip there, and had gotten a pretty good idea of what had gone on. They also found no radioactivity in the city, though they did find a trace of it about four miles downwind with an integrated dose of about two R; not enough to do any harm.

One evening there was a knock on my hotel room door and, remarkably, there was Stuart Harrison, erstwhile radiologist with Charlie Lauritsen at Caltech, erstwhile husband of Kitty Oppenheimer, and now a major in the Army Medical Corps. I don't know how he learned I was in Japan, let alone in the Dai Ichi Hotel; we must have been more famous than we knew. He was carrying a souvenir, a samurai sword. There's some sort of virus that goes along with military expeditions; everybody becomes a junk collec-

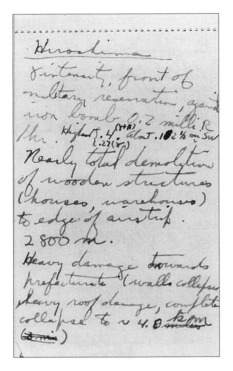

Fig. 6.16 Notes from visit to Hiroshima.

tor. I was also smitten, and brought home two altimeters and three compasses from Japanese military planes we found abandoned on the Hiroshima airstrip. I still use one of the altimeters as a barometer, rather than relying on TV weather predictions like everybody else.

We left Tokyo for Tinian and then home, flown back by the 509th. I had a little run-in with the pilot on the flight back to Tinian. I was hungry and no food was provided so, in a pet, I opened a five-pound tin of roast beef. The pilot was justifiably annoyed; he'd been saving the tin for the Japanese black market and he angrily explained to me how much it would have brought. I made another mistake at Tinian when I disembarked still wearing my Japanese woolens, and broiled overnight in the tropic heat.

We had a little trouble in San Francisco, where peacetime practices were now in effect. We had to go through Customs with our squashed gas cans, hunks of concrete, and charred crate, and through Immigration, and it turned out that Bill didn't have a passport. However, our other identifications so impressed the immigra-

Hiroshima Data from the Riken group: *11 Sept 45*

. Induced activity and neutron flux
From phosphorus beta-counts, they have estimated 10^{35} neutrons made in the expl osi
The activity drops to almost nothing at 800 meters from the center, has a radius
at half-ma ximum of about 400 meters, and fits roughly an inverse -square law
with an explosion height of 500 meters. Sulphur (n,alpha) results are similar,
but less accurate. The activity plot was quite sy metrical. They put the center
about 200 meters west of the aiming point, or a bit less.
An estimate of absorption in a wooden floor(bone activty on first and second floor
of the same house) gives a hydrogen scattering cross/section of about 2 barns. This
seems to me to be a pretty bad measurement.

. The result of blast damage was estimated to be equivalent to ten thousand tons of
high explosive.

. The maximum fission fragment activity was centered not on the blast but on an area
of a couple of square kilometers about four kilometers west of the zero point.
This region was surveyed first on about 20 to 25 August, and gave a maximum gamma
activity of about 1/2 milliroentgen per hour at that time. If we assume it took
of the order of five minutes for the cloud to settle there, we get an integrated
maximum dose from the ground of about 2 r units if none was lost by weathering.

4. Recommendations)
If we want any further data from Nishina, Dr. Serber should visit him and
discuss the situation. Nishina is anxious to bombard anyone with graphs and data,
evidently hoping to get a little confirmation of his own views.

For R Serber

P M orrison

Fig. 6.17 Hiroshima data taken from Japanese scientists by Philip Morrison and given to me.

tion official that he decided he could call Bill a British RAF officer and let him in.

We finally reached Los Alamos on the 15th of October. Charlotte filled me in a little on things that had happened during my three-month absence—about the big, high-spirited party after the news about Hiroshima, and about the gloom that had begun to settle in after Nagasaki. Charlotte also told me about bumming a ride to Hollywood with Shirley Barnett. The two had discovered a Navy plane flying out to Los Angeles, and spent the Labor Day weekend there. They visited our old friend Harry Kurnitz, now a successful script writer, and wound up at a party at the house of Ira Gerswhin, at which they heard Abe Burrows—a famous radio comedian at the time (who later became even more famous on Broadway, and who wrote the book for *Guys and Dolls*)—sit down at the piano and impro-vise songs, including one about the atomic bomb. This trip of Charlotte's and Shirley's was tracked by the FBI, and later Commo-

Fig. 6.18 Charlotte receiving the "E-Award" on behalf of the women of Los Alamos, October 1945. Oppenheimer is at the extreme left.

dore Parsons bawled out the pilot for letting women ride in a Navy plane. Later, an account of this trip wound up in Oppie's FBI file, as evidence of the kind of untrustworthy people he hung out with.

The day after I returned, the laboratory received an award for excellence, the so-called Army-Navy E-Award. The ceremony was a big event; General Groves was there, as was President Sproul of the University of California. Oppie accepted the E-Award for the laboratory, and Charlotte received it for the women of Los Alamos.

In the weeks immediately after my return, I gave some talks on the effects of the bomb, but I was able to begin working again on physics. Ed McMillan had come up with the idea of phase stability, which made synchrotrons possible, and I checked the mathematics for him. One day Charlotte noticed a Russian journal that contained an article, in English by Vladimir Veksler, with a very similar idea, and pointed it out to him. Luis Alvarez questioned whether bunching of the electrons would increase their radiation loss, but Julie Schwinger, at the MIT Radiation Lab, showed this didn't cause trouble. All this was closely connected to work Don Kerst and I had done on betatron theory.

That fall, the fall of 1945, there was a lot of frantic recruiting going on—people looking for jobs as well as universities looking for people. The University of Illinois offered me a full professorship; I'd been an associate professor when I took leave. But McMillan and Alvarez got after Ernest Lawrence to offer me a professorship in Berkeley at the Radiation Laboratory of the University of California, as head of the theoretical division there. Charlotte had never been pleased with smalltown life in Urbana, and I was glad too at the chance to be back in Berkeley and working in what was then the foremost accelerator lab in the world.

PART III

Peace Again

SEVEN

Berkeley, 1946–1951

Back in Berkeley at the beginning of January 1946, Charlotte and I had moved into one of the half-dozen apartments in the pueblo-like structure on Fish Ranch Road that we had lived in from 1936 to 1938. It was a crude but pleasant place, heated, Berkeley-style, by a huge fireplace in the living room. It had a nice terrace in front with a view, across Fish Ranch Road, of the Claremont Hotel gardens. We did most of our entertaining on that terrace. Once, when the younger Lauritsens were visiting us, Tommy and I were on the terrace when the smaller half of a dollar bill came floating down from out of nowhere. Tommy retrieved it and later got a big rise out of Marge, his wife, by using it to light a cigar. When we first moved in, the telephone was attached to a long, fifty-foot cord which was able to reach into every room in the apartment. However, when the first bill came, I discovered the phone company was charging us a dollar and a quarter a month for the extra-long cord. I called them and asked them to take it out, the phone repairman came, and thereafter we had a standard-length cord and no extra charge. About a year and a half later, we began having trouble with the phone bell, which worked only intermittently. We had another phone repairman up, who opened up the phone box—in those days, the bell was in a box on the baseboard—and discovered most of the fifty feet of phone cord coiled up and taped to the inside of the box. The tape had finally given way and the cord had fallen onto the bell. The original repairman had never bothered to remove the long cord, just rolled it up and tucked it out of sight. The new repairman took it out of the box, removed the tape, and asked us if we minded having a longer cord.

This time around, Charlotte got a cat—a kitten—rather than a dog. One day she missed it, and finally sighted it at the top of a telephone pole on the road, among the insulators and wires and too frightened to move. She called the fire department for help, but the fireman said, "Lady, we ain't been rescuing cats since before the war." When she protested he said, "Lady, you never seen a cat's skeleton in a tree, did you?" Charlotte finally realized it was a power pole, not a telephone pole, so she called the power company, which promptly sent a man who put on climbing irons, ascended the pole, and retrieved the cat.

I bought a surplus army jeep for transportation to work. The Rad Lab was now occupying the site on the hill above the campus where the 184-inch cyclotron had been started before the war. The site had been developed during the war, with new shops, labs, and office buildings, and the development continued with war's end. Our successes had convinced the military that scientists were a valuable national asset, to be nourished, and there was money for new labs, accelerators, research reactors, staff, postdocs, and graduate students. Not long after my return to Berkeley, the lab was visited by Secretary of War Ed Pauley. After we gave him a tour and were escorting him back to his car, Ernest mentioned that he could use an extra $2 million for the 184-inch cyclotron. Pauley assured him of the Army's full support for any such request. He got into his car and, as he leaned out the window for a farewell handshake, said, "By the way, Professor Lawrence, did you say two million or two billion?" It sounds apocryphal, but I heard it myself.

At the Rad Lab three new particle accelerators were being planned. Ed McMillan was designing a 300 MeV electron synchrotron, made possible by his and Veksler's theory of phase stability (a synchrotron is a greatly improved betatron, using an electric field for acceleration instead of a magnetic core). Ed's idea had also allowed redesign of the partly completed 184-inch cyclotron, for the phase-stability principle also made possible the synchrocyclotron, where the relativistic mass increase was compensated by changing the frequency of the rf (radio frequency) accelerating system, a great improvement on the original idea of using brute force with a 3 MeV energy boost per turn. Meanwhile, Luis Alvarez, in characteristic fashion, had posed to himself the question of what unique opportu-

Fig. 7.1 Ernest O. Lawrence (1901–1958), inventor of the cyclotron, the first particle accelerator to reach high energies, received a Ph.D. from Yale in 1925. He joined the University of California, Berkeley, in 1928, where he developed the cyclotron, giving the first scientific paper on it in 1930. Lawrence was awarded the Nobel Prize in Physics for development of the cyclotron in 1939. He was director of the Radiation Laboratory of the University of California from 1936 until his death in 1958. (Photo courtesy of Watson Davis, Science Service, and the University of California, Berkeley)

nities the end of the war would offer. He decided that a lot of surplus radar equipment would be on hand which could be used to power a linear accelerator, and so he planned and got approval to build a 30 MeV proton linac.

At the Rad Lab, I was director of the Theoretical Division. Our principal lair was a large room, about thirty by fifty feet, one corner of which was glassed in to create my office, while the rest was open, filled with desks and with a wall of blackboards. The postdocs complained about the lack of privacy, but pedagogically the arrangement worked out just fine; everyone, grad student or postdoc, was exposed perforce to the lively discussions at the blackboard. When I arrived in January 1946, two graduate students, Leslie Foldy and David Bohm, were at the Rad Lab. Ed had already given them the task of working out the theory of the synchrocyclotron. They wrote a classic paper on the subject, with only a little editorial help from Ed and me. Bohm and Foldy had many successors over the years, including Marvin Goldberger, Geoffrey Chew, Ken Watson, Richard Christian, Sid Fernbach, Ted Taylor, Ed Hart, Pierre Noyes, Keith Breuckner, Murray Lampert, and Robert Jastrow.

I still had some connection with weapons problems. In April 1946 I was put on a committee whose mandate was to evaluate the state of Edward Teller's program for the Super. We met in Los Alamos. As Edward and his group reported, it became apparent to me that at every point they were making the most optimistic assumption, and that no solid calculations had really been carried through. Near the end of the conference, Edward produced a version of the proposed committee report. I thought that in many places, particularly in the conclusion, it was overly optimistic. It said that the weapon was "practically certain" to work, and that there were just a few little details to be worked out. I went to Edward and proposed that we tone down some of the more outrageous statements, and together we went over the report and made it a little more realistic. I still thought it was very optimistic, but I had no objection to that—I had no desire to throw cold water on Edward's project and was all in favor of his proceeding with it as best he could, though I really didn't think there was any chance that that weapon would work as it was envisioned then. (The hydrogen weapon that Teller and Stanislaw Ulam came up with later was a very different kettle of fish.)

A couple of months later, in Berkeley, I happened to be in the library, and the librarian said, "By the way, a classified document has come in with your name on it—would you like to see it?" It was the final version of the conference report, "Report of Conference on the Super" (dated June 12, 1946). It was Edward's original report, with all the changes we had agreed on left out. That was unfortunate, because three years later, during the time of crisis which followed the explosion of an atomic bomb by the Russians, this report had considerable influence and people were misled about the likelihood of the Super working.

When Oppie left Los Alamos he went at first to Caltech and didn't return to teaching at Berkeley until the fall of 1946. As the year went on, he spent more and more time consulting in Washington. Although my appointment was as a research professor at the Radiation Laboratory, I would sub for him while he was gone, and as his absences became more frequent I gradually slipped back into teaching. In May 1947 he left Berkeley for good to become director of the Institute for Advanced Study in Princeton. I unobtrusively took over his teaching duties, and eventually I was given an appointment in the Physics Department in addition to that at the Rad Lab.

The Theoretical Division included an accelerator theory and design group, housed separately. I spent a good deal of my own time on accelerator theory. In fact, my first postwar paper was on this subject: Charlie Lauritsen's old student, Dick Crane, now at Michigan, had proposed and was building a "racetrack," a synchrotron with straight sections.[1] My 1946 paper, "Orbits of Particles in the Racetrack" (25), discussed the stability conditions for such a machine. The paper is of some interest because the problem has some similarities to that of alternating gradient theory, for Crane's machine involved alternating magnetic and field-free regions and was thus a special case of alternating gradients. Ernie Courant told me that when he was working out alternating gradient theory he found my paper helpful in figuring out how to solve the equations. I was also among the authors of the 1947 paper "Initial Performance of the 184-inch Cyclotron at the University of California" (26), by Brobeck, Lawrence, MacKensie, McMillan, Serber, Sewell, Simpson, and Thornton.

Besides Berkeley, other places where work was under way on synchrotrons were Brookhaven, Cornell, and the General Electric lab in Schenectady. One day Ernest called Ed McMillan and me into his office. He was upset by a call he'd just gotten from people at the GE lab who had told him they had done new calculations which showed that the synchrotron wouldn't work. GE had just built a computer, a mechanical analogue machine. John Blewett, then at GE, baptized it by giving it the equations for the phase oscillations of the synchrotron orbits. He was surprised and distressed when the computer showed the phase oscillations to be unstable, the amplitude of oscillation increasing with time. I laughed when I heard this and told Ernest I had solved the equations analytically and mathematics doesn't lie. Ernest was still concerned—it would be troublesome if a rumor of this reached Washington—and told Ed to go to Schenectady and look into it. After seeing their results, he asked the GE people to put the equation for sine x in the computer. Sure enough, the amplitude diverged. It turned out there was backlash in the gears.

I had one major failure in machine theory, in connection with the proton linac. Ed McMillan had proven a theorem that a proton moving parallel to the central axis of the machine would have no radial focusing if there were no charge in the beam area. To produce a charge for the radial focusing, a series of thin gold foils had been installed that the proton beam had to pass through. To avoid excessive scattering of the beam, these foils had to be extremely thin. While they worked, they were a big nuisance because they were difficult to make and were always breaking. Ed had proved that the first-order focusing effects vanished, and the obvious thing was to take a look at second-order effects—that is, the effects due to deviations from Ed's parallel path caused by the focusing system itself. It occurred to me to go to the library and see how focusing was done in electron microscopes. Over the course of the year, I resolved to do this at least half a dozen times. Once I got as far as the library door before being intercepted by somebody in the hall, and I never did get around to looking it up. If I had, and had thought about the problem a little, it's possible that I would have invented alternating gradient focusing before the Brookhaven gang did. It's a nice paradox that what I called second-order focusing—weak focusing—turned out to be the same thing as Brookhaven's strong focusing.

In Berkeley, Ed McMillan had come around with some ideas about machines, and he was talking about alternating gradients; I didn't look into what he had done in any detail. I said, "Oh, yeah, azimuthal variations of the fields—that's the Thomas cyclotron. Take a look at his paper." Neither of us pursued it.

The principle behind alternating gradient focusing is easily understood. Consider an optical analogy. Suppose one has a bunch of lenses of equal strength, half focusing and half defocusing. If packed alternately close together, there is no net focusing or defocusing effect—this case illustrates Ed's theorem. However, if the lenses are separated, the net effect is focusing, because the defocusing lens bends the light outwards, and when it strikes the next focusing lens, it is further from the center and therefore experiences a stronger focusing force. Similarly, passage through a focusing lens bends the light inward, and it hits the next defocusing lens closer to the center, where the defocusing force is weaker. The application of this principle to machine design would be discovered in the summer of 1952 at Brookhaven by Ernest Courant, Stan Livingston, and my old friend Hartland Snyder. (It had actually been discovered a few years earlier by a Greek electrical engineer, Nicholas Christofilos, but ignored, and even prior to that in the Thomas cyclotron.) John Blewett would then show how strongly focusing quadrupoles could replace the focusing foils in Luis's linac.

After the magnet for the 184-inch was completed and measurements were made on the shape of the magnetic field near the outer edge of the magnet, I used this data to predict the radii at which resonances would occur between the horizontal and vertical oscillations of the beam. At these points the amplitude of the vertical oscillations could increase and the beam be lost. I was able to predict the radius at which the beam would blow up, which was satisfactorily near the outer edge of the magnet.

While the 184-inch was being built, no preparations were being made for using it. The new high-energy regime was unfamiliar. I can remember a crisis at one point while the beam was being coaxed up to full energy; at a certain radius it suddenly disappeared. Everyone thought something had gone wrong and that the beam was lost. I realized, with a little shock of surprise, that nothing was wrong—what had happened was that the energy of the beam had become so high that its range was greater than the thickness of the probe being used as a Faraday cup to monitor it. The beam was passing right through the probe. All that was needed was a thicker probe.

The original beams were deuterons or alphas—the rf wasn't up to

Fig. 7.2 Alvarez and I at the 184-inch cyclotron, about 1947. (Photo courtesy of Gene Lester)

accelerating protons yet. Thinking about how to produce external beams, a nice mechanism occurred to me that would produce a high-energy neutron beam fairly narrow in both energy and angle: with a deuteron beam, a collision with a nucleus in the internal target could strip off the proton, allowing the neutron to continue ahead with half the deuteron energy. I first reported this idea at the July 1947 meeting of the American Physical Society (27, 28). Later in 1947 it was published with the title "The Production of High Energy Neutrons by Stripping" (29) ("stripping" was a Los Alamos term, referring to stripping an absorber from a uranium projectile by firing it through a hole in a metal plate), with the addition of comparisons with experimental neutron angular distributions measured by C. Helmholtz, Ed McMillan, and D. C. Sewell.[2] The agreement with theory was very good. Ted Taylor, who was a Berkeley grad student at the time, did most of the calculations. I've always regretted that Sid Dancoff's name didn't appear on the paper. Sid spent the spring semester in Berkeley on leave from Urbana and contributed significantly to the theory. When he returned to Urbana it was agreed that he would write up another mechanism for producing a neutron beam—disintegration of the deuteron by the nuclear Coulomb field—while I wrote up the stripping, but through my carelessness this turned into separate authorship.

On November 1, 1946, that notable night when the beam reached full radius and we were all celebrating, I remember Ernest Lawrence standing in the neutron beam with a Geiger counter held to his chest and clicking away, demonstrating with great glee his discovery that the counting rate was the same whether he faced the target or turned his back to it (the neutrons were passing right through his body). High energy took some getting used to. No detection apparatus had been prepared with which to do experiments. The first experimental results were obtained by Ed McMillan using nothing more complicated than radioactive foils and an ionization chamber.

The Rad Lab physicists, after years of war work, were feeling a little rusty on fundamentals, so they asked me, shortly after my return to Berkeley, to give a weekly lecture on things they should know or be interested in. These lectures began in the spring of 1946 and ran through part of 1947. A graduate student, Murray Lampert, took notes and printed his write-ups as Rad Lab reports under the title *Serber Says*. Two reports appeared, the first subtitled *About High Energy Processes and Nuclear Forces*, the second *About Mesotrons*. (We were still calling them "mesotrons" in 1947.) A demand for these reports developed all over the country, and the Rad Lab had to run off several printings. Murray Gell-Mann told me he had studied them at MIT when he was a student. My own copy of *About Mesotrons* was sent to me by Bill Nierenberg after he found it among his papers and bears the notation "Received May 17 1948: J. R. Dunning: Dept. of Physics: Columbia University."

A third volume of notes was compiled but never got published because Murray Lampert left the university. He wrote up the notes before he left and gave them to me in mimeographed form, and the reason they were never published was simply that I never got around to proofreading them and taking them to be printed. The content was interesting too: speculation on nuclear physics in the 100 MeV region. It may sound humdrum now, but, after predicting that nucleon-nucleon cross sections would fall off like $1/E$, I remember how surprised I was to realize that, in the energy regime we were

Fig. 7.3 Working on pions at Berkeley.

entering, the mean free path of a nucleon in nuclear matter became comparable to nuclear radii. Nuclei would be partially transparent, a notion quite different from the nuclear reaction theory to which we were accustomed.

I averaged over straight line trajectories through the nucleus to calculate absorption cross sections, and I introduced the impulse approximation, saying that the incident nucleon interacts with the nuclear nucleons as if they were free, except for a reduction in the scattering cross section due to the exclusion principle preventing small momentum transfers. This effect was later found to be basic in understanding the shell model. I calculated energy losses and cascades of secondary particles and, for the benefit of our radiochemists, distributions of residual nuclei. After the 184-inch worked and the predictions were seen to be being borne out, Ed McMillan prodded me into publishing a concise account of this work. Years later I learned that Charlotte had prodded Ed into prodding me; she knew I didn't like writing papers. In 1947 this one appeared in the *Physical Review*, entitled "Nuclear Reactions at High Energies" (30).

Early in 1947, MGM Studios released a movie about the Manhattan Project entitled *The Beginning or the End*, a pseudo-documentary starring Hume Cronyn as Oppenheimer. MGM invited the senior staff at the Berkeley Rad Lab to a preview of the film in San Francisco and since practically all of us had worked on the project we were curious. When the movie was over and the screening-room lights came up, it was clear we were expected to comment. There was a moment's awkward silence. Finally Ernest said, "Well, it wasn't really as bad as I expected!" Ernest was being uncharacteristically generous. All the scientists depicted in it were interchangeable—smart fellows with no personalities to speak of—bearing no relation to the people around me who were the purported subjects.

I flew East in June 1947 to attend the Shelter Island Conference, a conference of about twenty-five theoretical physicists that got its name from the fact that it was held on Shelter Island, in Peconic Bay near the far end of Long Island. The conferees assembled late one afternoon at the American Institute of Physics building near the UN where a bus was waiting to take them to Shelter Island. I didn't go on the bus: Willis Lamb was driving his own car and asked me to come with him, so I drove out with Willis and Rabi. We met the rest at a seafood restaurant in Greenport and heard that we had missed some fun. An escort of motorcycle police had met the bus on Long Island and escorted them in style to Greenport. The escort was a token of appreciation from a Suffolk County official who, as a Marine in the Pacific, had been slated to participate in the invasion of Japan. After dinner we took the ferry to Shelter Island and drove to the Ram's Head Inn. The next day Lamb and Rabi reported on the revolutionary measurements at the Columbia Radiation Laboratory: the Lamb shift and the anomalous g-value of the electron. There was much discussion, contributed notably to by Hendrik Kramers, on the problems of quantum electrodynamics. My own report on the first results from the 184-inch cyclotron, the first high-energy nuclear physics experiments, disappeared in the shadow of these great events.

In 1947 Cook, McMillan, Peterson. and Sewell began measuring the total cross sections and scattering angles of elements from lithium to uranium for 90 MeV neutrons produced by stripping from the 190 MeV deuteron beam: their results were published at the beginning of 1949.[3] To interpret them Sid Fernbach, Ted Taylor, and I introduced the optical model (31), in which the neutron wave passing through the nucleus is described by an index of refraction (or complex potential) to calculate the absorption, diffraction, and total cross sections. This was an absolute prediction: no parameters were adjusted. We used a measured 90 MeV n-p cross section and a nuclear radius deduced from 14 MeV neutron-scattering experiments. Theory and experiment agreed very well. At the 1949 Washington meeting of the American Physical Society I gave an invited paper on the optical model. Viki Weisskopf was chairman of the session. At the end of my talk, Karl Darrow got up and suggested the model should be called the "cloudy crystal ball," after a well-known department in the *New Yorker* magazine. Later, Viki and Herman Feshbach applied the model very successfully to lower-energy nuclear reactions.

During all of 1948, Hadley, Kelly, Leith, Segrè, Wiegand, and York were measuring the 90 MeV n-p scattering and Christian, Hart, and Noyes were working on the interpretation of the results. For the first time scattering was being done with neutrons of short enough wavelength to probe the shape of nuclear forces. The results showed a gratifyingly large departure from spherical symmetry. The most striking feature of the experimental curves was the approximate symmetry of the scattering about 90 degrees. Such a symmetry meant that the central forces were half ordinary and half coordinate exchange, or, alternatively stated, there were no central odd-ℓ forces. This was a surprising result, since it violated the "saturation" condition, which purported to show why nuclei didn't collapse until each nucleon interacted with all the others and gave a binding energy proportional to the square of the number of nucleons, rather than the first power, as actually observed.

Eugene Wigner visited Berkeley during this period and I told him how things stood. Later on, this combination of forces got to be called a "Serber force"; Eugene must have given it the name. The Serber force combination had another advantage; it minimized the total cross section by giving no odd wave scattering. This was important; although Dick Christian had no difficulty fitting the angular distribution, all the shaped forces he tried, constrained to fit the deuteron binding energy and quadrupole moment and the near-zero energy scattering, gave about 10 percent too large a total cross section.

I suggested to Dick that a repulsive core would reduce the cross section. The forces between nucleons—repulsive at small distances and attractive at large distances—would then be much like the forces between molecules in a liquid. In this way it would become possible to explain why the nucleus didn't collapse but instead maintained constant density as the number of nucleons increased, as a liquid does. Heisenberg had rejected this explanation in his classic 1932 paper because he considered such a complex interaction as characteristic of complex structures inappropriate for elementary particles (but of course nowadays one wouldn't call neutrons or protons elementary). But Dick reported, implausibly, that it spoiled his angular fit. I failed to follow it up. A little later, Bob Jastrow left the Institute for Advanced Study and joined our group, full of enthusiasm for the repulsive core, which he had thought of independently. He borrowed the stack of IBM punch cards containing Dick's program and in a day or two came up with a good fit to Segrè's data. So—a repulsive force at short distances explained the noncollapse of nuclei, and the shenanigans introduced by Heisenberg in 1932 were unnecessary. Nuclei *were* like liquids. Hadley et al. published in 1949,[4] Christian et al.[5] and Jastrow[6] in 1950.

The high energy of the Berkeley accelerators opened the possibility of an even more basic research program: study of the mesons Yukawa had said were responsible for nuclear forces. Lattes, Muirhead, Occhialini, and Powell discovered pi mesons in cosmic rays in 1947 just as the 184-inch got going and, of course, accelerator production of them became a major priority of the Rad Lab. Initially, the prospects weren't bright; we had available 190 MeV deuterons or 380 MeV alphas—only 95 MeV per nucleon. Accelerating protons, potentially to 380 MeV, required twice the rf frequency, which wasn't yet available. Teller pointed out that the momentum distribution of nucleons in the target nucleus, their "Fermi motion," would reduce the effective threshold energy. I remember at a meeting in the East Edward asking me if I believed the Berkeley cyclotron was producing pi mesons. I said I did, but the problem was to find them. Ed McMillan had designed and installed a target and nuclear emulsion plate holder. I estimated that, using the momentum distribution in both the alpha particle projectile and the target nucleus, we would still only get one pi disintegrating in the emulsion per million other tracks.

Gene Gardner was exposing, developing, and looking at the Ilford plates, without success until Cesare Lattes arrived in Berkeley. Cesare, coming from Cecil Powell's lab in England and experienced in technique and observation, promptly found pi decays, in February 1948. On March 9, a press conference on the artificial production of mesons was held at the lab. Afterwards, Gene, Ed, Lattes, and I retired to my office to relax, and a photographer from *Life* happened to wander by and snap our picture, which later wound up in Edward Steichen's *Family of Man*.

After the press conference, Lattes was feeling his oats; he felt he was someone important—but he didn't understand the realities of the Rad Lab. Ernest was a dictator, albeit a benevolent one. Norman Ramsey wrote Ernest congratulating him on the lab's achievement and saying he'd like to see one of the emulsions for himself. Ernest told Lattes to send a couple of plates to Norman. Lattes said no; it was his work and he didn't want any competitors horning in on it. Ernest fired him on the spot, and called a guard to escort him to the gate. A couple of days later, a chastened Cesare came to me for help. I took him to see Ernest, who greeted him as if nothing had happened, and Cesare went back to work. Needless to say, one of his first acts was to send Norman some plates.

Shortly thereafter, in March 1948, a second "Shelter Island" conference was held, this time at Pocono Manor, Pennsylvania. I was able to report Berkeley's first experiments with the pi mesons. Julie Schwinger was the star of this meeting, with his "manifest relativistic invariance" and his calculations of the anomalous magnetic moment of the electron and of the Lamb shift. I have a curious recollection of his talk, for it seems to me he spent a lot of time discussing setting up calculations on arbitrary space-time surfaces, a subject whose only real relevance may have been its effect on Julie's thinking. Then Feynman presented his now-famous Feynman diagrams. His reasoning was based on his own instinct rather than on any formal connection with the equations of electrodynamics. His casual approach seemed to annoy Bohr, who at one point got up and chastised Feynmen for ignoring the history of physics and making things up as he went along. At another point Feynmen expressed doubt about the existence of vacuum polarization, since he didn't see how to incorporate it into his scheme. After the meeting, he and I went out together, and I asked him if he had ever read any of the papers on vacuum polarization. He said he hadn't. I think that was one of Dick's strengths—when any question came up, he tried to understand it in his own way, not in the way someone else had thought of it.

Charlotte and I had some gay parties at our place on Fish Ranch Road, particularly at the time of events like the American Physical Society's meetings. At one of these, late in the evening, one of the guests enticed Charlotte off the terrace a few feet into the bordering woods and made a pass at her. When she resisted, he said, "Well, why not? Give me one good reason." Charlotte pointed to the ground and said, "Poison oak."

But Charlotte had been house-hunting on and off for a couple of years. House-hunting in Berkeley had its cautions. She was looking at houses on the hill, but the hills were notoriously unstable. One carried a marble and dropped it on the living room floor, and noticed how fast it rolled off into some corner of the room. In the spring of 1948, we finally settled on a house located on Santa Barbara Road. In the middle of the basement, it had a three-foot-square concrete pillar, from which steel rods about two inches in diameter stretched out to the four corners of the house. It probably wasn't a good investment, but the pillar gave us some confidence that the house would stay put. It was a typical California Spanish-style house: white plaster, red tile roof, rooms with high ceilings with wood beams. A big picture window in the living room looked out over San Francisco Bay—the city, the Golden Gate Bridge, Marin County, Mount Tamalpias. When Segrè saw it for the first time, he said, "Charlotte, you will pay for this view with your life." Segrè was convinced that San Francisco would be atom bombed within a few years; he himself moved out of Berkeley to Walnut Creek, twenty miles inland. When Rabi saw it he said, "The same view?"

One of our visitors was Dick Feynman who, we discovered, was occasionally a little manic-depressive. He would sometimes retreat to a room and close the door for long stretches of time. Once he shut himself in our basement with our dog Nicka, and emerged hours later claiming that Nicka understood eighty words of English.

I soon found there were responsibilities that go with house-owning. There was only a small patio cut into the hill in the back of the house out of the wind, so there wasn't much grass to cut. But on the downhill side, where there were sixty steps up from the street to the house, there were terrace gardens with about one hundred rose bushes that had to be sprayed once a week to keep the bugs off. We got a dog from the local guide-dogs-for-the-blind organization, a German Shepherd whose name was Nicka; they said she'd flunked

Fig. 7.4 Nicka, who according to Dick Feynman knew eighty words of English.

her final exam after two years of training as a seeing-eye dog. They just couldn't seem to keep her from chasing cats—but she was trained, as we soon found out. One day I was washing the car down on the street level, when Nicka came bounding down the stairs barking ferociously. She grabbed my pants leg and began dragging me towards the stairs. I finally got the idea, followed her up, and found Charlotte sitting on the ground immobilized with a sprained ankle. She had tumbled out of a tree that she had been trying to prune.

But any bucolic feeling of peace was soon interrupted. In the early summer of 1948, Robert Bacher, one of the atomic energy commissioners and an old friend, visited Berkeley to inform me that the AEC was going to put me through a security hearing. He assured me that they had no doubts about me, but said that they were forced into such actions to defend themselves from the (pre-McCarthy) hysteria in Washington, which included fears of Communists in the atomic energy program. Shortly thereafter, a Personnel Security Board was appointed to hear the case. The board was chaired by John Francis Neylan, the chairman of the Board of Regents of the University of California and William Randolph Hearst's attorney. The other members were Admiral Nimitz, who had been commander of U.S. Naval

Forces in the Pacific during the Second World War, and Kenyon A. Joyce, a retired major general. Late in July 1948, this board sent me a letter saying that an investigation into my "character, associations and loyalty" had raised questions about the continuance of my security clearance at the Radiation Lab, and listed nine items which, they said, comprised the "substance" of the questions. The letter invited my response to the items prior to the hearing.

I replied as best I could. None of the charges seemed very serious to me. The only thing they had directly against Charlotte was that she had been secretary of a medical aid committee for Spanish loyalists during the Spanish civil war, and during the Second World War she had been active in the British and Russian war relief committees. All the rest of it was guilt by association. They said that Charlotte's father and brother were Communists. I couldn't answer about their current activities, because we had only seen them half a dozen times on short visits in the previous fifteen years. In her father's case, it seemed unlikely to me, because he was an old-time Socialist, and I thought the Socialists and Communists disliked each other. As for her brother, I thought he had rather extreme political opinions and might well be a Communist sympathizer, though I doubted he would be a party member because one heard that the party demanded discipline of its members, and Milt wasn't one to submit to discipline. They also asked about our friendships with a number of people, including Mary Ellen Washburn, Oppie's prewar landlady in Berkeley, and David Bohm, who had been a graduate student in Berkeley when I returned after the war. I hadn't the slightest notion that there was anything subversive about either of them. They also listed Haakon Chevalier, whom we had known through Oppie in prewar Berkeley, and had seen once shortly after returning to Berkeley in 1946. At the time I had no reason to suspect there was anything questionable about this. The charges against him, which became notorious during the Oppenheimer hearings six years later, were not publicly known at the time—certainly not known to us.

The hearing took place on August 5, in Neylan's offices in San Francisco. The morning began not too auspiciously. When introduced to Admiral Nimitz, I said, "You know, Admiral, I was at your headquarters in Guam the day General MacArthur signed the peace treaty in Tokyo Bay." He replied, "That was the day *I* signed the peace treaty in Tokyo Bay." A real faux pas.

The setting reminded me of a movie court-martial, with three judges lined up behind a table, the accused in a chair facing them, prosecutors on the right—but with no defense attorneys or witnesses on the left. They asked me if I had brought along a lawyer or any character witnesses, and I replied that I didn't need any. I couldn't imagine there were any questions I wasn't willing to answer, so I couldn't imagine what useful advice a lawyer could possibly have given me. The board started out asking about the political activities of Charlotte's family, about which I knew practically nothing. Then they asked about Charlotte's prewar political activity, which I defended as admirable rather than reprehensible.

The hearing then developed in a way similar to what happened later on in the Oppenheimer hearing. They had listed nine charges in the letter to me—but this was only a tiny fraction of all the stuff they had in the files. They went on and on with many questions completely beyond the formal charges. They asked about friends like Frank Oppenheimer, and surprised me by saying he was a member of the Communist Party. They asked about students at the university, about whose political notions I had no idea at all. They asked about many people whose names I had never heard and whom I had no recollection of ever meeting. They asked me about Charlotte's trip to Los Angeles on Labor Day 1945, and her visit to Ira Gershwin's house in the company of Harry Kurnitz and Abe Burrows.

In one respect it was very different from the Oppenheimer hearings. It soon became apparent to me that there was not much substance in the files, and I think the same impression was shared by everyone involved. Whatever adversarial feeling there was at the beginning seemed to evaporate as the hearings went on; there was no equivalent of Roger Robb at the Oppenheimer hearings. By the time of the lunch break, I got the feeling that the board was not at all unfriendly, and at the beginning of the lunch break, Neylan advised me to call character witnesses. I rather foolishly demurred, saying I didn't like to disturb any of my colleagues. He then said he would call them himself, and asked whom I wanted him to call. I said Ernest Lawrence and Ed McMillan. They didn't succeed in finding Ed, but they located Luis and Ernest, who came for the afternoon session. Luis told them about how I was kicked off the plane to Nagasaki because of the missing parachute, and General Joyce in particular was much impressed by that story.

There was one interesting incident. At one point they spoke about Chevalier and an approach he had made to Oppie on behalf of a third party who wanted Oppie to pass on classified information to the Soviet Union; while Oppie had refused, he had also not told any security officials about it, either. I was asked what I would do if someone came to me and asked me to try to get Ernest to turn over some classified information. I replied I would try to stall them and tell Professor Lawrence. At that point there were sounds of consternation and disturbance in the audience, and I heard Ernest gasp, "Oh, no, Bob!" When the excitement died down, Neylan said, "We didn't really mean to set a trap for you, but when you said you would go tell Professor Lawrence, you meant you would report to the proper authority, Lawrence being in your mind the proper authority in this case." I said, "Yes, that's right. I don't know what I'd do if Professor Lawrence . . ." and then I trailed off in confusion. I think this little scene did more to convince the board of my unquestionable loyalty than anything else I could have said or done.

I don't recall getting any formal notification of the board's favorable judgment; I probably learned it from Ernest. I never saw the board's report. Later Oppie told me he had seen it and I had passed with glowing praise. But I had found the experience humiliating and frightening, and resented having been put through it.

By a curious coincidence, in our new house on Santa Barbara Road, our nearest neighbor turned out to be Admiral Nimitz, who lived in the house directly across the street. We met occasionally, mostly while I was out walking Nicka.

In the summer of 1948, the State Department had arranged with the American occupation authorities in Japan to let the Yukawas travel to the United States to work for a year with Oppie in Princeton. Charlotte and I met them in San Francisco when they arrived. Mrs. Yukawa's main desire was to see an American department store, so we took them to I Magnin's, the fanciest department store in the city. She seemed fascinated by all the things that were available. She was very visible, dressed in a kimono, and the Yukawas created quite a sensation. Japanese visitors to the United States were practically unknown up till that time, and wherever we stopped a small crowd collected, with antagonistic murmurs arising now and then. The next day, they visited the university at Berkeley, and I have a picture of them with Ernest and myself standing in front of the old radiation lab

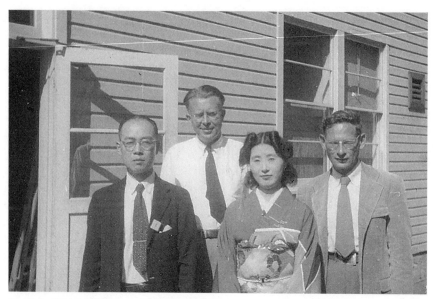

Fig. 7.5   Hideki Yukawa, Ernest Lawrence, Sumi Yukawa, and I in front of the old radiation laboratory on the Berkeley campus on the occasion of the Yukawas' first visit to the United States in 1948.

on the campus. Among my photographs, too, is a picture of Charlotte and Mrs. Yukawa, dressed in kimonos, in the patio of our house.

At the Rad Lab, many meson experiments were in progress. I went to the Solvay Conference in Brussels at the end of September 1948 to report on some of the work. The evening before the conference began, I ran into Paul Dirac in the hotel lobby, and he, Erwin Schrödinger, and I went to dinner at a nearby restaurant. During the meal, Schrödinger explained to us how he found the wave equation. The story has already been told by Dirac: how Schrödinger originally had the relativistic wave equation (the Klein-Gordon equation) and couldn't make things work; how he went to Paul Ehrenfest for advice; how Ehrenfest suggested he look at the nonrelativistic limit; and how that nonrelativistic limit turned out to be the correct equation.

The next morning, I arrived at the conference prepared to speak about the Berkeley experiments, and my talk was based on slides showing the apparatus and graphs of the results. I discovered that palaces, while pretty, are impractical: there was no way of curtaining the large windows, the room was impossible to darken, and my slides projected invisibly against the white screen. I was reduced to

Fig. 7.6 Solvay Conference, 1948.

trying to describe the instruments and graphs in words, and the talk was more or less a disaster. I was, however, able to report some conclusions I had drawn about the nature of the mesons. From the fact that pi-minus mesons, when captured by a nucleus, made stars in the nuclear emulsion, while mu mesons did not, I concluded that a neutrino carries off the energy in the case of the mu, but not in the case of the pi, which shows that the spin of the mu is $1/2$ and of the pi is either 0 or 1. I said the pi was decaying to mu plus neutrino and the mu to electron plus two neutrinos. This was the first suggestion, as far as I am aware, of the decay of the mu meson into an electron and two neutrinos.

At the end of my talk, Oppie raised a question about the coupling constant for mu decay and whether it was the same as that for beta decay. "Does the lifetime of the mu meson on the two neutrino hypothesis," he asked, "agree with the lifetime of the neutron"? I responded, "We have not finished the calculations, but I have seen no reason for a large change."[7] As a matter of fact, I had done the calculation the evening before leaving Berkeley but was unsure about a factor of the square root of two in the definition of the coupling constant. The fact that Oppie asked the question shows that the idea of a universal Fermi interaction was in the air.

Under more familiar and favorable conditions, I reported the same conclusions at an APS meeting at Berkeley; the abstract was in the May 1, 1949, issue of the *Physical Review*, under the title "The Spins of the Mesons" (32).

The third "Shelter Island" conference, at Oldstone on the Hudson in April of 1949, was timed just right for me to relate some more interesting Berkeley results. I reported there on the photoproduction of mesons, experimental and theoretical.

The explosion of the first Soviet atomic bomb, on August 29, 1949 (and announced by Truman a month later), caused a tremendous shock at the Rad Lab. Ernest was unnerved by the evidence of Russian technical ability; he had thought it would take them many more years to obtain a weapon. Teller immediately got in touch with him, pushing the idea of a crash program to develop the Super. He would have liked all nuclear physicists to give up their peacetime endeavors and concentrate on developing a thermonuclear weapon. I told Ernest that the Super wouldn't work; that Edward didn't know how to build a thermonuclear bomb. I urged him to talk to Hans Bethe to get an authoritative opinion. I don't believe he ever did. He fell on the idea of producing tritium and plutonium for a thermonuclear weapon, and—always the activist—he thought how he could do it himself. For the next couple of weeks he and Luis developed a plan to build heavy water reactors on the California coast, and even picked a site somewhere north of Berkeley. Luis was to be project director.

The General Advisory Committee of the Atomic Energy Commission had a meeting scheduled for October 28–30, in Washington. Ernest asked me to go and represent him at the meeting and present his proposal for building heavy water reactors. I'm sure that the reason for sending me was that he thought I would have influence with Oppenheimer, who was the chairman of the GAC. I didn't share Ernest's feeling that the country was in grave danger. I was more inclined to think that what did happen would happen—namely, that a stalemate would develop. Witness the fact that I bought a house facing San Francisco Bay while Segrè moved to Walnut Creek. But building production reactors, which could be used either to make plutonium or tritium, seemed a reasonable precaution. So I agreed to go. (I had no notion a paranoid arms race would develop.)

I went to Princeton on Thursday, the day before the meeting, and stayed overnight with the Oppenheimers. When I repeated to Oppie my opinion that Edward's Super wouldn't work, he warned me to be careful when speaking about that; while he said it was true that Edward's idea wasn't practical, some other idea might be, and he wanted me to be careful not to give a misleading impression. I know I presented Ernest's project to him, but I can't remember what his reaction was. He was much more interested in a communication he had just received from James Conant, concerning a proposal to add some advice on national policy to the technical report of the GAC. Conant used the word "genocide" in connection with the Super. He said it had no other military use and that the United States should not build such a weapon. It said that if the Russians did so and used it against us, we could very well retaliate with our stockpile of atomic bombs. To appreciate what was being said, it must be understood that the Super was conceived as a weapon with a minimum yield of megatons. In this it was very different from the Ulam-Teller H-bomb invented a few years later.

I was astonished; the East was evidently a completely different world from California. I had no idea that people like Conant and Oppenheimer would harbor any such ideas. At Berkeley they would have been unthinkable.

The next morning Oppie and I went by train down to Washington. That afternoon, at the AEC building, there was an unofficial preliminary meeting of the GAC members. At one point they called Bethe and me in to testify. Hans was asked about the Super, and if I recall correctly his opinion of it was no higher than mine. Then I talked of Ernest's concerns for increasing our production capacity for plutonium and tritium, and his proposal for the Berkeley Radiation Lab to build reactors for that purpose on the West Coast. Fermi then asked the obvious question: Why the Radiation Lab? Of all the labs in the country, it was the one with absolutely no experience in reactor design or operation. I answered that the proposal illustrated Ernest's great concern that new production facilities go ahead as rapidly as possible—a concern that was so great that he was willing to divert his own laboratory to the purpose—but if there were a better and quicker way of accomplishing that end, Ernest would be the first to be in favor of it.

The next morning the official meeting started. I met Luis in the lobby of the AEC building and we watched as the GAC members assembled, and later were impressed by the constellations of stars on the shoulders of the Joint Chiefs and other high-ranking officers going by to testify. At the lunch break Oppie came out and took Luis and me to a nearby restaurant. He told us something about how the meeting was going: the GAC was inclining against a high-priority Super program on both technical and moral grounds. I had been surprised to hear of Conant's position: Luis was appalled. He left for home that afternoon.

When I got back to Berkeley, I reported to Ernest what I had answered in reply to Fermi, and he said I had been exactly right. I returned gratefully to the physics of the lab. By 1949 we had high-energy protons in the 184-inch, and using them Bjorklund, Crandall, Moyer, and York found evidence of the pi-zero meson,[8] a discovery shortly confirmed by Steinberger, Panofsky, and Steller,[9] who photoproduced them using the newly operating 300 MeV electron synchrotron.

Further evidence on the character of the pi came quickly. The decay of the pi-zero into two photons proved the pi didn't have spin 1. For spin 0 pseudoscalar was indicated, rather than scalar, because of the tensor force between nucleons. Drawing on a number of Berkeley experiments, in a paper submitted in 1950 but published in 1951 with the title "The Capture of Pi-minus Mesons in Deuterium" (33), Keith Brueckner, Ken Watson, and I showed it was pseudoscalar. The argument depended on the observation by Panofsky, Aamodt, and Hadley[10] that the capture of pi-minus mesons in deuterium leads to the emission of two neutrons 70 percent of the time and two neutrons plus a gamma in 30 percent. Capture of a scalar meson from the K shell of the d-pi system is forbidden: the initial state is even with J=1, while the only J=1 state of two neutrons is odd. Using detailed balancing and measured cross sections for the inverse processes of meson production in nucleon-nucleon collisions and photoproduction of mesons, we could prove that the capture rate from orbits of higher angular momentum was too slow for scalar mesons, and that all seemed in agreement for pseudoscalars. This paper is notable because, it was reported to me, Fermi said it was the best-written paper ever to come out of Berkeley, which I took to be less a compliment to our paper than a comment on Berkeley literacy. Oppie's papers were often highly abbreviated—he wrote letters when he should have written articles.

In an extension of this work we published a paper, "The Interaction of Pi Mesons with Nuclear Matter" (34), interpreting a number of experiments dealing with the interaction of pi mesons and light nuclei. But this was the last paper I submitted from Berkeley; there was trouble in Eden.

Early in 1950, after almost a year of controversy, the Regents of the University of California reaffirmed a requirement that all faculty and employees of the university take an oath swearing that they were not a Communist and were loyal to the United States. This created a tremendous uproar on the campus, and poisoned the atmosphere of the university for years. There was anger in the debates, people were at odds with each other, and it got even worse when the Regents actually began to fire people—even tenured professors—for not signing. Gian Carlo Wick said that he had been coerced into taking an oath once before in Italy, where he had to swear loyalty to Mussolini; he said he'd regretted it ever since and wasn't going to make the same mistake twice. Gian Carlo was fired, so was Geoffrey Chew, and Wolfgang "Pief" Panofsky was so upset that the affair played a role in his leaving Berkeley for Stanford. I was unhappy, but I didn't take it so seriously that I wouldn't sign. However, I was really offended when my friends and colleagues were dismissed. I stayed on for a year or so longer. When Wick left I inherited his people. One of them was T. D. Lee, who after he received his Ph.D. had come out to Berkeley to work with Gian Carlo at Fermi's suggestion. Wick left just as he arrived, so T. D. began to work for me. Formally, that is: in fact, T. D. didn't need much direction.

But then another unfortunate element appeared, in the form of a growing political opposition between Ernest Lawrence and Robert Oppenheimer. Ernest had come from a fairly liberal North Dakota background, but as time went on his opinions gradually grew more conservative, while Oppie was a liberal Democrat. As Ernest began to oppose Oppie's policies more and more in Washington, I grew more uncomfortable. As a member of Ernest's laboratory, I was supposed to be one of Ernest's boys and support his views of the world, with little latitude for independence. For instance, when Ernest asked me to go to Washington to represent him on the Lawrence-Alvarez reactor project, I had no real option to refuse. And just around that time, Ernest was turning the effort of the laboratory over to the calutron project—the Materials Testing Accelerator—and I was finding myself having to spend a good deal of time going to meetings about calutrons.

Early in 1951, Rabi came to Berkeley and for the second time played his avuncular role in a way that affected my career. He told me in effect, "You have to choose between Ernest and Oppie." Of course there wasn't any doubt about which way I'd choose, but it was still very difficult. I was very unhappy about it because I was fond of Ernest and he'd been a good friend. It was painful, but Rabi offered me a job at Columbia and I went.

It was an excrutiating decision. Charlotte and I had a nice house— the only house we would ever own, it turned out. And while at Berkeley I was the boss of the whole division, at Columbia I would be, for a while at least, just another professor. I did my best to disguise the fact that my departure was due to the conflict between Ernest and Oppie. We told everyone it was for family reasons— which wasn't altogether untrue, since Charlotte's parents had moved to New York City in the interim. Charlotte's sister Madi, who had become widowed in 1945, became the employee, lover, and then wife of George Ross, a New York theatrical press agent, and Morris and Jenny had moved to New York to be with her. Another tempting part of the offer was the proximity of Columbia to Brookhaven National Laboratory, whose director was my old Wisconsin colleague, Leland Haworth.

Fortunately, it turned out that Columbia and Brookhaven were lively places for a physicist to be.

EIGHT

Columbia and Brookhaven, 1951–1967

Rabi not only persuaded me to come to Columbia; he, or rather his wife, Helen, also found Charlotte and me a place to live. Their own university-owned apartment was a block away from the campus on Riverside Drive, overlooking Riverside Park and the Hudson River. Helen had noticed her neighbors across the hall preparing to move, and got Rab to pull the necessary strings to reserve the place for us. We moved in that summer, and I've lived there ever since.

When we arrived in July, no one was around Pupin, the Columbia physics building; everyone was away for the summer. I spent the summer, until the beginning of classes, at the Nevis Laboratory, an adjunct of Columbia's Physics Department about twenty miles up the Hudson on an estate that had belonged to Alexander Hamilton—the island of Nevis in the West Indies was his birthplace. The Nevis lab housed a 300 MeV cyclotron built by Eugene Booth and James Rainwater that became operational early in 1950 and also served as a staging ground for experiments at Brookhaven. No one ever took the summer off in those days; one always worked somewhere under some contract. There were no other theorists at Nevis, and it was a little lonely. Jim Rainwater was the only one I remember talking to much. But it was always a problem talking to Jim; he had an original mind, and he looked at physics problems in a different way than anyone else. He used his own language to talk about them in terms nobody else used. When you had a discussion with him, you spent the first half hour figuring out what his words meant and what his point of view was before you could understand what he was talking about.

My office in Pupin Laboratory, which housed Columbia's Physics Department, I owed to George Pegram. Pegram, a physicist who had taught at Columbia since just after the turn of the century, had recently retired as Columbia's vice president to become a special adviser to its president (then Dwight D. Eisenhower), but still had his old Pupin office as well as his administrative office in Low Library. Polykarp Kusch, the Physics Department chairman and one of its Nobel laureates, asked him to give up the Pupin office so I could have it, and Pegram agreed. (The only place and time in history when office space was not a problem was Berkeley right after the oath controversy.) That office had been the scene of a historic meeting on December 16, 1940, at which Pegram, Ernest Lawrence, Enrico Fermi, and Emilio Segrè met to discuss the possibility of discovering whether plutonium was fissionable. There, they planned a cyclotron irradiation of uranium to produce enough plutonium to measure its fission cross section.

Willis Lamb was leaving Pupin just as I was moving in. He was on his way to Yale. I had seen him only a couple of weeks before in Berkeley. I drove him up the hill to the Rad Lab, and when we got to my office he said, "You're crazy to be leaving Berkeley for Columbia." I said, "Why?" He said, "Here you have a parking place with your name on it." Now in Pupin I was sitting in his office while he packed the contents of his desk. As he finished he presented me with a ten-inch porcelain crucible that had sat on his desk as an ashtray large enough so that visitors wouldn't miss it. Willis himself didn't smoke. I still have it; when I quit smoking it transformed into a kitchen utensil.

Just before leaving Berkeley I had received a letter from Kusch asking me to teach an undergraduate course, "Introduction to Mathematical Physics." He included a synopsis of the course as it had been given, and I was appalled. First of all, I had never taught an undergraduate class in my life, outside of the lab sections I had as a graduate student (and I've never taught one since). Department heads who knew me, like Loomis and Birge, never asked me to teach undergraduates. Perhaps they thought I was too highbrow, but, more concretely, I spoke in a very low voice, even in the classroom. Graduate students would put up with it, but I didn't think undergraduates would. While I was a grad student at Wisconsin my mentor, Professor Van Vleck, had sent me to the Speech Department to see if anything could be done to correct my speech. On my first day there the professor had me read a poem to his class. After he had tutored me for most of a semester, he impressed the class by having me recite the poem in a loud, clear, resonant voice. However, it didn't take, I never used that voice for anything else. Second, the course evidently had been designed about thirty years previously, and was mostly concerned with the mathematical physics of the nineteenth century. I was leery, in a new place, about making radical changes in what was evidently a traditional course. I wrote to Kusch protesting the assignment. He obliged, and switched his program around. Yukawa, who was then a professor at Columbia, had been scheduled to teach quantum mechanics. Kusch gave me quantum mechanics instead and Yukawa an advanced quantum mechanics course. Yukawa's English was good, but I bet he wouldn't have done much better with undergraduates than I would.

That fall, I began teaching at Columbia. My first Ph.D. student there was Leon Cooper, whose dissertation was concerned with the interpretation of Jim Rainwater's experiments on mu-meson atoms. Leon shared the 1972 Nobel Prize in physics with John Schrieffer and John Bardeen for their theory of superconductivity.

Nevis was supported by the Office of Naval Research. That had an unexpected consequence in the spring of 1953 when I applied through Nevis for travel funds to attend an international theoretical physics conference in Japan, the first major postwar physics conference in that country, scheduled for the end of 1953. I was astonished to hear from Rainwater that the request was refused. Jim had heard that Naval Intelligence had vetoed my request. Those were the days of Joseph McCarthy, and Naval Intelligence was afraid of adverse criticism if it came to light that I had been supported by the Navy. This in spite of my clearance by the Personnel Security Board. I was really offended, and my resentment played a significant part in my later refusal to become a consultant to the Department of Defense during the Vietnam War. Charlotte took it as confirmation of the wisdom of her postwar policy of avoiding any political action, for fear of such repercussions.

The Columbia physics staff was extraordinary. Rabi and Yukawa were Nobel laureates, and in the ensuing years nine other Nobel prizes were to be won by staff members. Theorists, in addition to Yukawa, were Norman Knoll and Henry Foley, each of whom had contributed significantly to developments in quantum electrodynamics during the previous years. There was also Shirley Quimby, who had just retired and whose Math Methods course I had been asked to take over. Quimby had two extracurricular distinctions: he was Measurer of the New York Yacht Club and had the responsibility of measuring the America's Cup contenders, and he was the nephew of the famous entertainer, Blackstone the Magician. When his uncle died, Quimby inherited his paraphernalia and for years after, at department social gatherings, he amused with his magic tricks. The department also included two women professors—Lucy Hayner, who ran the student laboratories, and the distinguished nuclear physicist Chien Shiung Wu, who won the first Wolf Prize in 1978. And in 1953, T. D. Lee joined the department.

But at first I found the new job a little lonely. It wasn't just that I was no longer director of a theoretical division. There was a different atmosphere at Columbia, due of course to Rabi, who believed experimenters should do their own theory. He always complained there were too many theorists. I missed the Berkeley give-and-take. But as Nevis grew, it drifted towards the Berkeley model. Rabi foresaw it and at first resisted providing any office space at Nevis in a futile effort to keep the professors working at Nevis tied to Pupin. I was pushed towards the Berkeley mode by Jim Rainwater, who balked at Nevis's paying summer salaries to theoretical graduate students, saying all I had to do to get money of my own was to apply for it. So I did, and found myself Principal Investigator of a Columbia theoretical group supported by the Atomic Energy Commission. Within a few years our AEC contract was helping support a half dozen or more grad students, a half dozen postdocs who were at Columbia on government fellowships, a couple of assistant professors, and whichever professors needed travel money or a summer salary. During subsequent years the Physics Department also added some senior theorists: Quinn Luttinger arrived in 1960, Mal Ruderman came for a year on an NSF fellowship in 1952, returning for good in 1969, and Gian Carlo Wick, who had been at the University of Pittsburgh since being fired at Berkeley, came on board in 1965.

In 1957 Lee and Chen Ning Yang won the Nobel Prize. In addition to being the youngest person to win the prize, I think T. D. holds another speed record. One night in 1960, about eleven, my phone rang. It was T. D., calling from Princeton (that year, T. D. had left Columbia for the Institute for Advanced Study). He told me he wanted to come back to Columbia. I immediately broke the news to Poly Kusch, the Physics Department chairman, who also lived at 450 Riverside in the apartment on the third floor directly below mine. Poly conferred with George Frankel, the dean of Arts and Sciences, who lived on the ninth floor, and fifteen minutes after T. D.'s call, I called him back to say it was settled. A full professorship arranged in fifteen minutes.

Shortly after his arrival in 1953, T. D. initiated "Chinese lunch," which became a Columbia tradition. At noon on Mondays, before the Theoretical Seminar, a group of about fifteen would meet at Pupin to escort the visiting seminar speaker to lunch at a local Chinese restaurant, usually the Moon Palace at 113th and Broadway or, until it closed, the Shanghai at 125th. T. D. would order many dishes, and the audience would barely get back in time for the seminar at 2:10. That time was also a Columbia tradition, picked, originally, to allow Norman Knoll to repark his car for the city's alternate-side-of-the-street parking deadline. And a couple of times a year, for some special occasion or visitor, there would be a full Chinese banquet, usually arranged by Luke Yuan and attended by the whole department, including wives.

Through ignorance I ran afoul of a more venerable tradition, the "colloquium" every Friday evening at 7:30. Shortly after my arrival I was asked to run it, and as soon as my selection for the job became known, I was petitioned by most of the members of the department to move up the time to 4:30 in the afternoon so they could have Friday evening free. The change seemed reasonable to me and I announced it. I heard later that Rabi was outraged. The 7:30 colloquium was a Columbia institution, he said. He pointed out that it was attended faithfully by physicists from the surrounding schools—NYU, CUNY, the Stevens Institute of Technology—who couldn't get there in the afternoon. But the deed was done and the time was never changed back—in fact it was later moved up to 2:10.

The first research I did at Columbia was on a problem of Rabi's. He asked me to look critically at a method he proposed for doing atomic beam measurements on excited states of atoms. So for most of

Fig. 8.1   In Gertrude Goldhaber's office at Brookhaven; T. D. Lee is at above, left.
(Photo by Gordon Parks, *Life* magazine; copyright © Time, Inc.)

my first year at Columbia, I left meson theory and nuclear physics
completely and worked on his atomic beam problem. When it was
completed, I turned the manuscript over to Poly Kusch, and it circulated for a while around the Columbia Radiation Laboratory.
Eventually it disappeared, and I didn't have the will to rewrite it.
Sixteen years later, in 1968, someone, while cleaning out an office at
the Radiation Lab, discovered the manuscript, and Kusch thought it
was still of sufficient interest to be published. He sent it to the *Annals
of Physics*, and it was published under the title, "The Theory of
Atomic Beam Double Resonance Spectroscopy" (50).

   Soon after arriving in New York, I paid a visit to Brookhaven,
where my old friend and colleague from Wisconsin and Urbana, Lee
Haworth, was now the laboratory's director. Lee asked me to become
a consultant, and from then on I spent one day a week at Brookhaven.

For almost the next twenty years, I would drive out in the morning and back in the evening every Thursday, a trip each way of about an hour and a half. When I first started, one traveled a maze of three-lane and two-lane highways to get there. For the next few years, they were building the Long Island Expressway, which every year extended a few miles further out on the island. As it extended, the traffic grew with it, and the time it took to get to Brookhaven was invariant—an hour and a half before the expressway, an hour and a half afterwards.

There were a number of old friends on the staff at Brookhaven: Hartland Snyder, one of Oppie's students, was working on accelerator theory for the Cosmotron; Ken Green, who had helped Jerry Kruger build the Urbana cyclotron, was constructing the Cosmotron; Maurice and Trude Goldhaber, nuclear physicists from Urbana, were now on the Brookhaven staff. In the beginning I was the only high-energy theorist at Brookhaven and was kept quite busy on that one day a week. There was always a line of people wanting advice or instruction or just to talk about the news in physics. I also contributed to the ongoing theoretical seminar.

The first paper I published from the East Coast was in collaboration with Hartland Snyder, a command performance ordered by Sam Goudsmit, who was editor of the *Physical Review* as well as chairman of the Physics Department at Brookhaven. Sam was getting bored by a controversy in the letters column of the *Physical Review* concerning the standard way of calculating beta decay energies by taking the energy difference between the initial and final atoms. Some people were claiming that this way was incorrect, because the atomic electrons in the final atom didn't have time to readjust before the beta particle escaped. Hartland and I settled the issue by showing that there was such an effect, but that these people had much overestimated it (35).

In the summer of 1952, we moved out to Brookhaven. The summer sessions there were memorable. There were accommodations for about fifty families on the laboratory site; one-story, bungalow-type shacks, rather primitive apartments, not too unlike the accommodations at Los Alamos. Physicists from all the Eastern universities would come and bring their families. After work we would go swimming at Smith Point or Westhampton beach, and on weekends a picnic was usually held at Westhampton beach. In those days, when one crossed the bridge to the barrier island at Westhampton and turned

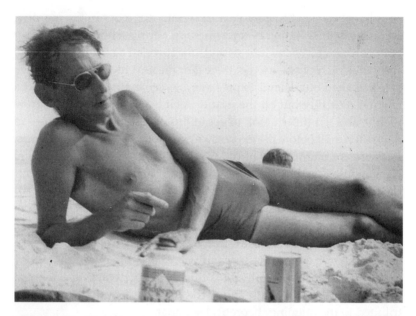

Fig. 8.2 Working at Westhampton Beach.
Fig. 8.3 Enrico Fermi with Priscilla Duffield, at Westhampton Beach.

right, there was nothing between the bridge and the public beach a few miles west but a tennis club at the bridge end, a snack bar at the beach end, and a couple of old Victorian mansions in the middle. It was not long before this beachfront became built up solidly with elaborate summer houses. Even at beach picnics, physicists can't forget physics, and one of my memories is of T. D. Lee writing equations with a stick in the wet, smooth sand near the water's edge. Among the pictures I took at the beach in those days is one of Rab in a bathing suit, as well as a picture of Fermi talking to Priscilla Green Duffield, the Universal Executive Secretary. Priscilla had been Ernest Lawrence's executive secretary when he started building the electromagnetic separators in Berkeley, and then became Oppie's executive secretary at Los Alamos, where she married Bob Duffield. After the war they went to San Diego where she was Roger Ravell's executive secretary at the Scripps Oceanographic Institute. After that they went to Argonne National Laboratory, and the only reason she wasn't the executive secretary there was that her husband was the director, so instead she became Bob Wilson's executive secretary in the early days of Fermilab.

One day, in the early fall of 1952, I got a message that Lee Haworth wanted to see me. I went around to his office, and he handed me a sheet of paper which he said contained an idea for an improved accelerator, which had been developed by Ernie Courant, Stan Livingston, and Hartland Snyder. He asked me to take a look at it and see if I thought it was correct. It was—it was the alternating gradient focusing idea. Of course I felt foolish, I should have thought of it myself. I missed it in connection with the focusing in Luis Alvarez's linac. When I finished the "racetrack" paper (25), I had a feeling that I hadn't pushed the work to its logical conclusion, which might have led me to alternating gradient focusing. And a little later in Berkeley, Ed McMillan had brought me some sheets of paper in which he discussed a synchrotron with azimuthal field variations, but instead of thinking about it I referred him to L. H. Thomas's paper on the Thomas cyclotron—another anticipation of the idea. Of course, Brookhaven immediately began to plan a new machine, the Alternating Gradient Synchrotron, which would be a successor to the Cosmotron.

In December 1952, Brookhaven dedicated the Cosmotron, the first BeV accelerator. It was a big event covered by the press and TV. At

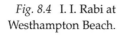

Fig. 8.4 I. I. Rabi at Westhampton Beach.

the culmination of the ceremony, I was standing on the balcony, looking down at the two thousand–ton magnet ring which circled the floor, when a TV reporter tapped my shoulder and asked, "When does it move?" That evening, we celebrated at a dinner in the gymnasium. It was a cheerful event, and we drank more than usual. When the after-dinner speaker—Detlev Bronk, president of Johns Hopkins and of the National Academy of Sciences, got up to speak, I wasn't paying much attention until I heard him say something about Canada. I began to listen, and realized he wasn't talking about Brookhaven or the Cosmotron at all. It turned out that his secretary had handed him the wrong speech, and he was reading one due to be delivered a couple of days later at some event in Canada. Det must have realized the mistake early on, but didn't act a bit flustered and just went ahead and finished it.

About three in the morning of April 13, 1954, I was awakened by a phone call; a voice on the other end identified itself as belonging to Samuel Silverman, a lawyer in Lloyd Garrison's office. He told me that Oppie was going to have a loyalty hearing, that the news would break the following morning, and that Garrison wished him to convey a message to me, namely, that I shouldn't try to communicate at all with Oppie until the end of the hearing. He said the Serber name was connected with Oppie's left-wing associations. He said that Oppie's phone would be tapped and he would be under surveillance, and any communications between us could only do harm. This news was a tremendous shock, but there was a big headline in the *New York Times* the next morning: "Dr. Oppenheimer Suspended by A.E.C. in Security Review."

I abided by that lawyer's request, and didn't try to get in touch with Oppie. Later on, Kitty told me that Oppie knew nothing about it, that it was all Garrison's idea. She said I should have paid no attention to Silverman. Like the great majority of physicists, I was saddened and outraged by the conclusions of the Personnel Security Board. Oppie's downfall was engineered by the Air Force, which wanted to get rid of advisers to the government who had advocated the DEW line, the early-warning radar system in Canada, and also a project involving tactical weapons for Europe, which promised to divert billions of dollars from the Strategic Air Command to the Army. Oppie also had some personal enemies: AEC chairman Lewis Strauss, and Edward Teller.

When the transcript of the proceedings was published, I was appalled by the tactics of Roger Robb, the counsel for the board. The hearings were supposed to be nonconfrontational and for the purpose of establishing the truth. But Robb acted like the prosecutor in a criminal trial, whose goal was to gain a conviction. Shortly after the publication of the transcript, Charlotte received a letter of apology from Norman Ramsey, who thought he had not sufficiently defended her when Robb asked him about her "communist tendencies." Norman felt that Robb had set him up and misled him.

Oppie and I never discussed the hearings. He was a sad man afterwards, and his spirit was broken. After the war, he had built his life around his advisory positions with the government. When these were ended, he seemed to feel he had nothing to live for. Bram (Abraham) Pais told me that, the morning he first read about the proposed hearings in the *New York Times*, he rushed to tell Einstein about it. He was astonished when Einstein burst into laughter. Einstein said, "The trouble with Oppenheimer is that he loves a woman who doesn't love him—the United States government." Oppie was still director of the Institute for Advanced Study, but Rabi remarked that that was something he could run with his left hand. He spent his time writing and lecturing in a half-hearted way.

As soon as we got settled in New York, Charlotte began looking around for something to do. George Ross, Madi's new husband, was the theatrical press agent for shows like Cole Porter's *Kiss Me Kate*. Through George, Charlotte got a job as secretary to Eddie Choate, a manager and producer of Broadway plays. On the first day of her theatrical career the first thing she was asked to do was, "Get me a dog that can make a stage entrance by itself!" She solved the problem, and later made quite a reputation for herself as a solver of demands like "Get me Helen Hayes," by a simple technique: she looked in the phone book, a notion that apparently never would have occurred to anyone else in the theatrical world. In the case of the dog she found an outfit called "Animals Anonymous" which could supply one. When Charlotte visited the apartment of the woman who ran it, there was a llama in the living room and a leopard had to be shooed off the couch. A second reason for Charlotte's success was that she wasn't an aspiring actress, playwright, or scene designer. For the next dozen years, she served as production assistant for a number of producers.

Jack [Jacob] and Marian Javits lived in our building at 450 Riverside Drive. This was before he became senator, when he was attorney general for the State of New York. One morning shortly after she had started work, Charlotte met the Javitses in the elevator going down, and they offered her a lift to work. Their chauffeur dropped them off first, and then took Charlotte the rest of the way to her office—making a smashing impression on the doormen of her office building: one of the secretaries (as far as they were concerned) arriving in a chauffeured limousine with license plate NY 3. One morning Leonard Lyon's popular gossip column in the *New York Post*, "Lyons Den," had an item: seen last night at Sardi's, Nobel Prize-winning physicist Robert Serber dining with the Javitses. But I suppose our highest point of social visibility was attending the opening night of *My Fair Lady*. Everyone else in the theater was recognizable. Charlotte knew the production assistants and secretaries of all the other Broadway producers and they exchanged passes. We got to see nearly every play that opened on Broadway.

One of the people Charlotte worked for was Huntington Hartford, the A&P heir, who wrote a play based on *Wuthering Heights* and hired Errol Flynn to play the lead. They had quite a job keeping Flynn sober enough to play anything. The opening night party was held in Hartford's apartment overlooking the East River and the United Nations and was enlivened by a dozen models from Hartford's modeling agency. The party was going fine until the arrival of the reviews in the early editions of the morning papers.

Charlotte also worked for a long time for Elaine Perry, whose family owned the *Denver Post*, and whose mother, Antoinette Perry, founded the Tony Awards. Their biggest hit was *Anastasia*, about a young woman who is possibly the daughter of the last Czar of Russia, and who had supposedly escaped execution. It had a moment of Grand Theater, the "Recognition Scene" between the girl claiming to be the Grand Duchess Anastasia, played by Vivica Lindfors, and the Dowager Empress, played by Eugenie Leontovich. In the later movie version the parts were played by Ingrid Bergman and Helen Hayes. At the time, Vivica Lindfors, a Swedish actress, was one of the most popular movie stars in Europe. On election night in 1956 (Eisenhower versus Adlai Stevenson), Charlotte and I tried giving a mixed party, physicists and theatrical people, including Vivica. As might have been predicted, the two sides didn't mix. And spirits were damped by the election returns.

Another producer she worked for was Heila Stoddard, an actress who was the star of a popular TV afternoon soap opera. Our maid, Winnie, used to watch the soap opera every afternoon while ironing. Winnie was completely confused one night when Charlotte had Heila over to dinner, which Winnie served. She was utterly unable to separate the TV character from the reality. One of Heila's plays starred Shirley MacLaine, at an early stage of her career before she became a Hollywood star. The first time I saw her she was being dragged down a flight of stairs by a pair of enormous borzois.

Bill Rarita, who had been in my group at Los Alamos, was now a professor at Brooklyn College, and he and I collaborated on a paper about proton-proton scattering with large momentum transfer, which was published in 1955 (36). The interesting thing about that paper was that we treated the proton not as an elementary particle but as something with structure, a radical departure from the usual treatment.

In the fall of 1954, Abraham Pais took a sabbatical from the Institute for Advanced Study and spent it at Columbia. Bram had come to this country immediately after the war, having survived the German occupation in Holland by the skin of his teeth. I had first met him at the 1947 Shelter Island conference, and later we became good friends. Bram and I discussed the results being obtained at Brookhaven on Cosmotron experiments with K mesons, and wrote a paper together in which we drew conclusions about the interaction of K mesons with nucleons from scattering, and the behavior of K mesons in nuclei (37). We continued our collaboration after Bram returned to the Institute, and a year and a half later published a paper entitled "Strong Coupling" (39). At the time, an interest in theoretical physics was focused on formal properties of scattering amplitudes, not restricted by the magnitude of the coupling constant. There were many efforts to extend the results of perturbation theory calculations of scattering to larger coupling constants. Bram and I thought it would be instructive to start at the other end, with large coupling constants, and work down to the intermediate range. We therefore embarked on a more critical and careful discussion of the strong coupling theory than had been attempted before. In our first paper we discussed the charged scalar theory, which had certain advantages of simplicity. This was followed, two years later, by a paper in which we treated the symmetrical pseudoscalar theory (40).

But there was something about the strong coupling papers that bothered me. When I examined the feeling, I realized that there was a gnawing question that went all the way back to my paper with Dancoff in 1942 (23). When Sid and I first did the strong coupling theory, we separated the meson field into the self-field (that is, the Yukawa field) and the free field part. However, Sid then went to work with Pauli and they did something different. They separated it into the source function and a different free field. When Sid came to write up the work he and I had done, he didn't do it the way we had originally set it up, but followed the Pauli-Dancoff procedure. The thing that bothered me was that the original way of doing it seemed to make physical sense, while the Pauli-Dancoff method didn't.

I got to work with a graduate student, Harry Nickle, and we developed the original Serber-Dancoff method, and as a matter of fact were able to show that for the charged scalar case it gave a much more controllable and tractable procedure than the Pauli-Dancoff one (41). In the course of our work, we picked up an error in the paper Pais and I had written—which was fortunate for me, because while Harry and I were working I happened to run into Pauli at the Institute for Advanced Study. He immediately jumped me, and said that he didn't believe a statement Pais and I had made in the charged scalar paper, that there were stable isobars no matter how small the coupling constant. I was able to admit our mistake and tell him that I now had a correct treatment, in which isobars weren't stable until the square of the coupling constant was greater than two.

While Berkeley had invented the high-energy physics lab, it really wasn't a user's lab and was in effect for the Berkeley staff. The user's lab was invented at Brookhaven, but the labor pangs were not easy. At first the lab invited proposals for experiments on the Cosmotron from prospective users, and the idea was that they would be done pretty much in the order in which they were proposed. In no time at all there was a couple of years' backlog and endless problems of overlapping proposals. Though there was a technical committee to advise George Collins, the chairman of the Cosmotron Department, it concerned itself only with logistics. George had to decide the experimental program, and soon everyone was mad at George and he was being driven crazy. By 1955 Lee Haworth decided something had to be done, and that the right thing to do was to create a Program Advisory Committee. You can't imagine what a radical proposal this seemed at the time. The idea that somebody would judge whether your experiment was worse than someone else's—it was insulting, insufferable, and probably unconstitutional. Haworth convened a Cosmotron users' meeting, but no one at Brookhaven wanted to stand up and propose such a committee. Lee thought it would be good to get an outsider—a theorist, even—to propose the idea, so there could be no conflict of interest. I was elected to break the news to the users, which I did and got plenty of verbal brickbats but survived. I paid a price. I was on the Program Advisory Committee for a good many years, first under Leland and then under Maurice Goldhaber.

Meanwhile, much experimental work with mesons was going on at the Cosmotron. Important results were the discovery, in 1956, of the long-lived K meson by Leon Lederman and Willy Chinowsky and the experiments on K regeneration by Oreste Piccioni. I remember pointing out to Piccioni, in a discussion of regeneration, that one must expect a small difference between the masses of K-long and K-short. Oreste has always been very conscientious in giving me credit for this observation. George Collins set up the first counter array to try and measure large momentum transfer events. I urged Collins to look for jets. The jet idea was nothing new: Bremstrahlung shows a jet phenomenon involving photons. I thought the same kind of argument would apply to particle fields. Collins did try to look for jets, but his apparatus did not have the sensitivity needed to find them.

The long interval between the two papers that Bram and I wrote on strong coupling was due to the fact that I took a sabbatical leave from Columbia during the academic year 1957–58. I received an invitation from Homi Bhabha to spend several weeks teaching at the Tata institute in Bombay. On the basis of that, Charlotte and I planned an around-the-world trip. Charlotte only flew first class; she had an idea that first class was safer. She also always believed, for equally valid reasons, in staying in the best and most expensive hotels. We left in May, with the idea of spending most of the summer vacationing in Europe. In England we rented a car and toured around for a few weeks, then had it shipped to France. In Paris, Maurice Levy had invited me to give some lectures at the École Normale, and we stayed there for a while. After checking into our Paris hotel, we got in the elevator and ran into Glenn Seaborg, then chairman of the Atomic Energy Commission. "Why, Bob," he said, "I didn't know you were here!" "I'm not," I replied. The elevator door opened and Glenn got out, ending the conversation. Charlotte thought it a funny exchange, but it seemed to me clear enough. Glenn meant that he didn't know I was on the American delegation to whatever conference he was attending and I said I wasn't.

After Paris we drove south and stayed a couple of weeks on the French Riviera, where I first saw a bikini, thence to Italy where we saw the usual sights and were greatly impressed. In the Sistine Chapel we ran into Viki Weisskopf, who was also taking a sabbatical.

From Rome we flew to Istanbul and spent a few days sightseeing there. One day we took a ferry trip up the Bosphorus almost to the Black Sea, near the Russian border. Charlotte swore that there were intelligence agents from at least three different agencies watching us on that trip. I thought she was probably imagining things. But the next day when we got out of the elevator on the seventh floor of the Istanbul Hilton and entered our room, Charlotte said, "Did you notice that little man sitting on the bench opposite the elevator door? He's been there every time we come out of the elevator." I said it must be a coincidence. She said, "No, there's never any man on any other floor." The next time we came into our room, Charlotte closed the door, waited a minute, then opened it again and peered out. She said she saw that man entering the room next to ours. On the day we were leaving Istanbul, Charlotte got a bad case of the trots, and her life was complicated by the fact that the checkout time was a couple of hours earlier than the time we had to leave for the airport; she sat around uncomfortably in the lobby for a while—and pointed out that the little man was also in the lobby, not on the seventh floor.

From Istanbul we flew to Ankara, where the physicists Fesar Gursey and his wife Suha had invited me to lecture at the Middle East Technical University. Fesar was then spending half the year as a professor at Yale and the other half at the Middle East Technical University. A short time before, while at Yale, he had received orders from the Turkish government to return to Turkey to complete his military service, from which he had been excused while a student. Fesar knew that Rabi was American ambassador to the United Nation's Science Committee and was about to attend a meeting of that body. He asked Rab to intercede with the Turkish ambassador and try to get the order rescinded. But apparently Rabi overdid it. A couple of months later Fesar received another communication from the Turkish government, canceling the military service but ordering him to report immediately in Istanbul to take up his new duties as Chief Science Advisor to the Turkish Air Force.

Fesar's invitation had been seconded by Erdal Inönü, also a professor at the Middle East Technical University, who had received his degree at Princeton under Wigner. Erdal's father, Ismet Inönü, was a very distinguished man. He had been a general in the Turkish army and Ataturk's chief lieutenant during the war for Turkish independence after World War I. When the Turkish Republic was founded in 1923, Kemal Ataturk became president and Ismet Inönü prime minister. It seems that proper names were not customary under the Ottoman empire but were among the reforms introduced by the Republic. An act of parliament gave Mustafa Ketal the name Ataturk, or "Father of the Turks," and Erdal's father the name "Inönü" after the place where he won a battle during the war of independence. When Ataturk died in 1938, Erdal's father became president of

Turkey, a post he held until 1950, when he paid the price for introducing free elections in Turkey by losing the first one. At the time we visited, he was out of office, leading the Opposition. He briefly became prime minister again in 1964.

I gave my lectures at the Middle East Technical University, which I found to be a lively and progressive place. It was a new institution. The buildings had been designed by a young Turkish architect who did a marvelous-appearing job—big wide halls, rough-finished concrete walls, and vistas to the outside wherever you looked. But it turned out there were some disadvantages to this modern rough-finished concrete construction. When classes were changing it sounded as if many herds of elephants were rushing back and forth, and it was completely impossible to hear anything in the classroom. And the students rapidly discovered that the big halls with wide vistas were very good locations for ping-pong tables, which also detracted from the quiet university atmosphere.

One day Erdal invited us to come to his father's house and have tea with his mother. A chauffeured limousine picked us up at our hotel and created quite a stir. When we got back the bell boys, room clerks, and waiters practically got down on their knees. Ismet Inönü was a legendary figure, and an invitation to his house was much more prestigious than, say, an invitation to the White House. After we got back, we were VIPs at that hotel for the rest of our stay. A year later, T. D. Lee was also invited to the Technical University and stayed at that hotel on our recommendation. During his stay, in 1960, a military coup overthrew the Turkish government, and from his window T. D. witnessed fighter planes strafing the Parliament Building, just a few blocks away. He said it reminded him of his old days in China during the war.

One day the Inönüs, Gurseys, and Serbers took a couple of cars and made a trip out into the Anatolian Plain to visit the old Hittite ruins. We stopped at several villages on the way, and at each one a crowd gathered around our cars when we got out. They recognized Erdal, who apparently looked very much like his father. It seems the peasants were mostly members of the opposing Democratic Party, angry members it seemed to us and rather threatening. But there was no overt trouble and we finally got to the Hittite ruins. There was no one else around. At one point there was an unattended watchman's shack near a pile of clay tablets covered with inscriptions. Erdal told us a watchman had been installed because the local peasants had been carrying off the tablets for building material. Nearby was a fortress with dirt walls and a fifty-foot tunnel leading into its interior. The Hittites obviously didn't know how to make an arch, and to create a tunnel had dug a trench, leaned together two rows of rock slabs each eight or nine feet high, and then filled in earth around them. The tunnel was dark, and Charlotte, who was claustrophobic and hated anything of this sort, at first refused to enter. It was pointed out to her that the tunnel had stood up for the last six thousand years and was unlikely to collapse that very afternoon. Reluctantly, she finally took a few steps in, then squealed and backed out as she detected, through open-toed sandals, the recent presence of sheep.

While we were in Ankara I was invited to give a talk and inspect the Turkish atomic energy installation near Istanbul. I flew down there and gave my talk and was shown the swimming pool reactor, which was in the process of being built. At one point chanting began from a nearby mosque and the workmen at the edge of the nuclear reactor pool put down their prayer rugs, got on their knees, and knelt towards Mecca, which struck me as highly anachronistic. As we were about to leave, the director of the installation said, "Now just a minute. I have to reimburse you for your airfare." He turned to his male secretary, who disappeared into the office and came back apparently quite flustered and whispered to his chief. I didn't know what was going on, but the director said, "Oh, we'll get the check in a few minutes. Meanwhile, let's go to lunch." We got into a limousine and were driven to a resort on the Sea of Marmora which looked like a Hollywood movie set. We lunched on an elegant open-air terrace on the water's edge, with speedboats plowing back and forth just off-shore towing girls on water skiis. Our lunch surely cost several times as much as the airfare. Then, without any further word of explanation, we were driven into Istanbul, where the director asked me to follow him into a huge block of a building. The first floor was a large open space filled with hundreds of clerks scribbling away in ledgers. The director led me upstairs where he introduced me to a very distinguished-looking gentleman who turned out to be the finance minister of Turkey. It seems that only the finance minister himself could get any funds in less than three weeks, and he personally escorted me to the cashier's window and signed the necessary papers to get my reimbursement.

After Ankara our next stop was Israel, where I was to lecture at the Weizmann Institute in Rehovot. We were impressed on arriving to see the guard at the gate of the Institute carrying a machine gun. Shortly after we arrived, we went on a sight-seeing trip down to the Dead Sea. Coming back, the old rickety bus we were on began to slip its clutch on the long uphill grade. Every few minutes the bus had to stop until the clutch cooled off. As a result, we were an hour late for a dinner at Rehovot given in our honor by Chiam Lieb and Leah Pekoris (he was a physicist famous for his description of the transmission of seismic waves through the earth). Our hosts had been worried, because that same bus excursion a week before had been fired on by terrorists and a couple of people had been killed. After I finished lecturing, the Weizmann people kindly presented us with a young physicist who had a wife and car and took us on a several-day sightseeing trip around the country.

Our next stop after Israel was to be Bombay. We had to go back to Istanbul because we couldn't fly from Israel over any Arab country. When we returned to the Istanbul Hilton, we got a room on a different floor from the one we had earlier—but when the elevator door opened, there was the same little man, sitting on the bench opposite the elevator. "Oh, you are looking much better now!" he said to Charlotte.

On our way to Bombay we stopped briefly at midnight at the airport in Karachi, where we saw a newspaper bearing the news that Lee and Yang had won the Nobel Prize in physics, and we cabled our congratulations. In Bombay, Bhabha had arranged a room for us in the Taj Mahal Hotel. Our room was in the old part of the hotel, which was not air-conditioned. Charlotte thought that was fine and that we'd simply enjoy the local atmosphere—until she opened the window and an insect with about a twelve-inch wingspan flew right into the room, something like a dragonfly. That made her want to change rooms, but as no air-conditioned room was available we had to stay put. The next thing she did was take a bath. While I was in the bedroom, the door suddenly flew open, a hotel servant entered with an armful of towels, and without a word he marched through the bedroom, into the bathroom, changed the towels while paying no attention to Charlotte in the bathtub, and marched out again.

Our room in the Taj Mahal Hotel overlooked the harbor, and directly beneath us was the Royal Yacht Club. Shortly after we arrived came a weather report of an approaching storm, and we were amazed to see how the Yacht Club handled it. Several hundred men appeared who waded into the water and surrounded each boat—as many as could get around each hull—and then simply lifted the boats out of the water and carried them up on land. The whole procedure couldn't have been done in an American yard, with one or two hoists, in anything like the time; here, a dozen crews were working at once. We were constantly amazed at the effect of the cheap labor. The first evening before dinner there was a knock on the door; three men were there and asked if they could make up the beds for the night. We said we weren't ready yet. An hour later, when we left, the three men were sitting in the hallway in front of our door, having waited all that time. Another time we took a taxi to see the Red Fort and dismissed the driver, which seemed to surprise him. When we came back out, we had trouble finding another cab. We shortly learned that, with cabs charging fifteen cents an hour to wait, one always kept them for the return trip.

The poverty was hard to adjust to. One night Charlotte and I went to a movie to see an Indian film, couldn't find a taxi on our return. During a walk of a couple of blocks to a taxi stand, we had to step over the bodies of half-naked sleeping homeless people, and by the time we reached the stand Charlotte was practically hysterical.

The Tata Institute, where I was to give a course in nuclear physics, was located way out of town on a promontory next to the Arabian Sea. While it was a very impressive building, it wasn't suitably air-conditioned and the hot moist air played havoc with the apparatus in the laboratories. I believe it was financed by the Tata family, the Rockefellers of India. Bhabha himself was a Tata and, during our stay in Bombay, invited us to a reception of the Tatas. I remember a conversation with a pleasant gentleman who asked me about the best jewelers in New York: he had a diamond necklace worth about a quarter of a million dollars and was finding the property taxes on it onerous.

Our old friend Bernard Peters, who had been Oppie's student at Berkeley in the prewar days, was a professor at the Institute, and before I'd left New York he'd invited me to lecture on low-energy nuclear physics. But Bernard was a cosmic ray physicist, and it turned out that what he meant by low energy was something in the neighborhood of a hundred million volts, while I was prepared to talk about one volt. I had to improvise. Bernard told us of some of the difficulties of teaching in India. High-caste Hindu students at first thought it was beneath them to clean up after themselves in the laboratory. And when they did begin to absorb the scientific outlook, it made for serious trouble with their families as they abandoned the taboos of their culture.

From India we visited the caves at Adjanta and saw the paintings there; we also visited Alara, where the temples were hewn out of solid rock. We flew to Delhi, rented car and driver, and took a several-day trip around that area. At the Taj Mahal we were surprised to find that we were the only foreigners there that day, plus maybe a dozen Indian tourists. We visited Fatehpur Sikri, the Mogul capital which was abandoned in the early sixteenth century, though it looked as if the people had left only yesterday—no desecration, no graffiti, no initials. Very un-Western. Another visit was to the Amber Palace, where we were carried up the road to the entrance on an elephant's back, followed by a troupe of musicians playing native instruments. This time we were the only people there.

From Bombay we went to Bangkok as sightseers, to visit the temples. We also made a side trip to Cambodia to see Angkor Wat. Charlotte was assured that we would be taken in a four-engined plane, but it was only just—the engines looked like converted outboard motors. We landed near the temples in a grass field, the kind one had to chase the goats off before landing. There was a customs shack, however, and the customs officers practically rolled on the floor and died laughing when they realized Charlotte had paid the official rate for Cambodian money. No one had ever done that before. She had been told better, but for some reason was terrified of all officialdom and always went strictly by the book. After an overnight stay at a local hotel, we were taken around the ruins by a guide who picked up a rock and knocked a piece of frieze off a wall and handed it to us. In the afternoon we wandered off by ourselves in complete isolation among ruins overgrown by jungle. Not quite complete isolation—when we were climbing down a steep set of stairs from an old temple we ran into another couple coming up. "Why, Bob and Charlotte Serber," they said, "Remember us? We are So-and-So from the English Department at Urbana!"

At our next stop, Hong Kong, we stayed at the elegant Peninsula Hotel, whose only shortcoming was that there was water in the bathroom for only two hours a day. In the course of the usual Hong Kong shopping, I bought a handsome, two-foot model of a Chinese junk for $7.50. They assured me that they could safely mail it to the United States, which I didn't believe, but it seemed worth the risk for $7.50. When the model finally arrived, it did so inside, not the sturdy crate which I expected, but a flimsy crate made of thin slabs of wood. The crate had taken a beating, and was twisted and half-broken, but the model wasn't damaged at all.

We celebrated that New Year's Eve in the Peninsula's dining room. Not only were the women in elegant costume, so were the men as well, half of whom wore fancy-dress military uniform while the other, civilian half was equally gaudy in brightly colored brocaded jackets. From Hong Kong we flew to Japan, where we spent a few days visiting the Yukawas in Kyoto. In Tokyo we met Sagane, the physicist whom Alvarez, Morrison, and I had written from Tinian.

The trip back home across the Pacific was on an American plane— the first we'd been on in the whole year of traveling through Europe and the Orient. Our first impression was that the American personnel and hostesses were incredibly rude; we had gotten used to a different standard of service. We stopped over for a couple of days in Honolulu, where I took a lesson in riding standing on a surfboard. Pretty soon I was successful, and rode all the way in to shore on a not-very-big wave. The trouble was getting off. When I tried, the board flipped up and gave me a bloody lip.

In July of 1959, in a gesture of international friendship, the Rochester Conference was held in Kiev. Our first Russian experience was on the Russian plane that was to take us from Helsinki to Leningrad. When we were seated, Charlotte discovered that her seat-belt was broken. She called over the hostess, who said "Nichevo!" ("Oh, it's nothing!"). Later the hostess handed out oranges, and we refused ours, much to the amazement of the other passengers. We soon found out that oranges were generally an unobtainable luxury in the north of Russia.

Russia had only recently been open to visits by Americans, and travel was strictly regulated by Intourist. However, our scientific meeting was a special kind of event that somehow managed to slip through the Intourist net. After a couple of hours of struggling through immigration and customs in Leningrad, we found that no provision at all had been made for what to do with us. We got a taxi and had considerable difficulty communicating with the driver, who was disgusted with us that we couldn't speak Yiddish, but who took us to a hotel called the "October." While no one there spoke English, they did speak French, which Charlotte knew well enough to manage. After sightseeing in Leningrad, we flew to Moscow, where Intourist caught up with us and installed us in the National Hotel— an old hotel, and much more interesting than the Metropolitan, the modern tourist hotel a short distance away. Our large room boasted heavy brocade curtains, a grand piano, a window looking out on the Kremlin, and the dust of years. Charlotte was impressed by the Russian hotel's management of the guest's requests. The woman at the desk had a large notebook in which whatever one asked for— transportation, air or rail tickets, ballet or theater tickets—was written in pencil in the next line, and everything actually happened the way it was supposed to.

Sputnik and the beginning of the Russian satellite program were still recent, and we found a toy consisting of a replica of a spaceship with a plastic dog's head, on a spring, sticking out the window. It was also the year of the first American fair in Moscow, when Vice President Richard Nixon had his famous altercation with Kruschev. While touring the fair, we ran into an exhibition with photographs from Edward Steichen's book *The Family of Man*. One of the photographs had my picture in it, along with Ed McMillan, Cesare Lattes, and Gene Gardner. While we were looking at it, some Russians noticed the resemblance, and we were soon surrounded by a big crowd.

The scientific meeting at Kiev was well organized. The only thing the organizers neglected to take into account was the fact that, at lunchtime, no Russian restaurant serves food in anything less than three hours. At the time, theoretical physics was governed by a fad for bootstrapping and democracy. Most people no longer believed in quantum field theory, no doubt because of its technical and calculational difficulty, and were willing to settle for very partial descriptions of nature which related a number of structures, from protons to uranium nuclei, on an equal footing. I thought it was all nonsense. When Geoff Chew, the leading exponent of democracy, got up to speak, I went out to have a cigarette in the lobby. Lev Landau followed me out, and said teasingly, "Serber, I thought you had better taste—do you still believe in point field theory?"

One day of the conference was devoted to an excursion boat ride down the Dnieper River. We stopped at a remote spot for lunch, a swim, and a visit to a memorial for a Ukrainian poet, Shevchenko, after whom our excursion boat was named. His memorial was way up on a hillside, and to get to it one had to climb up a stairway consisting of several hundred steps. Halfway up we found a level resting space with a kiosk which, surprisingly, sold books. Even more surprisingly, the books were all textbooks—some of them in English, including English-language physics textbooks. From Kiev, we also took a side trip to Odessa on the Black Sea to enjoy the balmy weather and Black Sea resorts. There were lots of oranges in Odessa.

We returned to New York that summer (1959), which we divided between Brookhaven and sailing on Long Island Sound. A couple of years before, my colleague Henry Foley had introduced us to the joys of sailing on the Sound in a thirty-five-foot schooner. After several trips with Henry and his wife Peg, we bought a boat of our own, a thirty-foot gaff-rigged Alden sloop—Alden had been a leading American yacht designer—built in 1926. The boat, which was called *Whitecap*, was something of an antique, in the way of antique cars, and attracted a great deal of attention whenever we sailed into a harbor. Shortly before Christmas in 1959, we received an invitation from Stuart Harrison, Charlie Lauritsen's ex-radiologist and Kitty's ex-husband, to go sailing with him in the Virgin Islands. He had received an inheritance in England, and used it to buy a fifty-foot steel-hulled ketch, named *August Moon*. We met Stuart and his new wife, Helen, in Charlotte Amalie, St. Thomas. One of our first chores was to provision the boat. We went to a local supermarket, and whom should we run into right off but Oppie and Kitty. It was a little embarrassing all around. Oppie and Kitty had just built a house on the beach in St. John. There was no supermarket on St. John and its inhabitants had to take the ferry to St. Thomas to do their shopping.

Sailing was great in the Virgin Islands in those days. If another boat pulled into the same harbor you were occupying you had a feeling of outrage. One night we anchored in Trellis Bay in the British Virgins. Stuart had arranged for us to have dinner at Marina Key, and he dressed up in very English fashion, yachtsman's cap and blazer, and we left in the dinghy with Stuart intent on making a grand entrance. This was spoiled when, approaching the dock, his outboard didn't turn off when the switch was thrown, and we hit it head

Fig. 8.5   Murray Gell-Mann (1929—) entered Yale University at the age of fifteen, received his B.S. in physics in 1948, and his Ph.D. from MIT in 1951. In 1952, while at the Institute for Nuclear Studies at the University of Chicago, he invented the concept of "strangeness," a property that would help explain the decays of so-called "strange" particles. In 1962, he (and, independently, Yuval Ne'eman) invented an important scheme for classifying particles, while in 1964 he introduced the idea of quarks. He won the Nobel Prize for physics in 1969.

on, tumbling everyone into the bottom of the dinghy. The next morning we left Trellis Bay and headed north. A little earlier, I had looked at the chart and saw there was a reef to the north that we needed to bear east to avoid. However, before leaving New York, Charlotte had given me a strict lecture on behavior, to the effect that Stuart was the captain and I was crew, and not to interfere with or criticize his navigation. Remembering this instruction, instead of warning Stuart, I ran forward to the bow where, sure enough, I saw coral heads rapidly approaching underwater. I waved frantically, but before Stuart caught on to the meaning of my flapping arms we crashed into the coral. We managed to kedge off without much difficulty, and an overboard dive showed that the steel hull hadn't sustained any damage aside from loss of some anti-fouling paint.

We returned to New York in January 1960 and resumed the familiar pattern of life at Columbia.

In March 1963, Murray Gell-Mann was scheduled to give a colloquium at Columbia. A couple of weeks before his talk, we asked Gian Carlo Wick to give a colloquium to provide background for Murray's talk. Gian Carlo spoke mostly about the symmetry group known as SU(3), which would figure prominently in Murray's discussion of the particle system. Symmetries play an important role in analyzing quantum mechanical systems. To illustrate, consider a much simpler system, an atom containing a number of electrons. Since the electrons all have the same mass and charge, the wave equation for the system is unchanged if one interchanges the coordinates of any pair of electrons. All the possible interchanges form a group of transformations called the symmetric permutation group. A consequence of the wave equation being unchanged by a group of transformations is that a number of states have the same energy, and such states are said to belong to an irreducible representation of the group. If there are n states of the same energy, we say we have an n dimensional irreducible representation. The atomic system has another symmetry group called SU(2), the "2" referring to the fact that the electron can exist in two spin states, spin up or spin down, and the wave equation does not depend on whether the spin is up or down. When there are two symmetry groups, the states that form irreducible representations of one group also belong to an irreducible representation of the other. As a graduate student with Van Vleck, I had done a lot of work on energy levels of atoms and molecules, and published some papers on the permutation group. I knew the irreducible representations of SU(2) and used this to find the irreducible representation of the permutation group.

In 1961 Gell-Mann had discovered that the baryon system of particles—the particles including protons and neutrons and heavier ones that could decay into protons and neutrons—showed evidence of SU(3) symmetry. The spin $1/2$ baryons had the isotopic spin and strangeness characteristics that could be associated with an eight-dimensional representation of SU(3), and the spin $3/2$ baryons formed a ten-dimensional representation of SU(3). Only nine of the ten were known at the time, and Murray predicted the energy and other properties of the tenth, which was soon discovered at Brookhaven by Nick Samios. The spin 0 mesons belonged to the eight-dimensional representation.

The analysis of the SU(3) group was not a simple problem. It had been worked out by mathematician Herman Weyl, and in his talk Gian Carlo told us about Weyl's results for the irreducible representations of SU(3). Thinking about it the next day, it occurred to me that I could find the irreducible representations by a low-brow method, using a generalization of what I had done for atoms. Instead of SU(2), the group appropriate for electrons with spin up or down, I could get an SU(3) group by considering a particle—now called a "quark"— which could exist in three states. For the n electron system I had used the known irreducible representations of SU(2) to find those of the permutation group. Now, for an n quark system, I could reverse the process and use the known irreducible representations of the permutation group to find those of SU(3). When I did this for the three-quark system, I found the eight-dimensional and ten-dimensional representations of SU(3), just those found for the baryons, while the quark-antiquark system gave the eight-dimensional representation of the mesons. The suggestion was immediate: the baryons and mesons were not themselves elementary particles but were made of quarks—the baryons of three quarks, the mesons of quark and anti-quark.

Before Murray's colloquium, I took him to lunch at Columbia's Faculty Club and explained this idea to him. He asked what the charges of my particles were, which was something I hadn't looked at. He got out a pencil and on a paper napkin figured it out in a couple of minutes. The charges would be $+ 2/3$ or $-1/3$ proton charges—an appalling result. During the colloquium Murray mentioned the idea and it was discussed at coffee afterwards. I remember that the name for the particles came up. Bacqui Beg, professor of physics at Rockefeller University, says he recalls that the name *quark* was produced at that time—that Murray had said that the existence of such a particle would be a strange quirk of nature, and quirk was jokingly transformed into quark.

A day or two later, it occurred to me that while the quarks' fractional charges were strange, their magnetic moments would not be. The magnetic moments depended on the ratio of charge to mass. In the nucleon the quark would have an effective mass one-third of the nucleon mass, so the one-thirds would cancel out in the ratio and the quark would have integral nuclear magnetic moments. A simple calculation gave the result that the proton would have three nucleon magnetic moments and the neutron would have minus two, values quite close to the observed ones. That convinced me of the correctness of the quark theory. At that point, I should have published; but I never got around to it. Bacqui Beg suggested to me that the reason was that the idea seemed so obvious to me that I thought it must be familiar to the experts in the field. However, it was news to Murray, and some time later he told Marvin Goldberger that he had never thought of it.

In 1963 I published a paper (43) in which I asserted that the observed large momentum transfer p-p elastic scattering fell off with an inverse sixth power law, a mistake due to uncritical treatment of the data. Around this time I wrote a number of papers on scattering theory. Charlie Townes was at Columbia and asked me about what quantum effects would be on a laser. What role would the quantum fluctuations have, especially in the start-up of the laser? He and I collaborated in a paper on the subject (42). And I wrote another report for the Columbia Radiation Lab entitled "Response of Superconducting Surfaces to Electromagnetic Waves" (48).

We were at Brookhaven in the summer of 1963 when Kitty and Oppie came to spend a few days there. It was the first good opportunity we had to be with them in some time. Oppie was to give the fifth Pegram lecture, a lecture series established in honor of George Pegram (whose office I now occupied in Pupin), who had done much to get Brookhaven off the ground. In 1946 Pegram had written a letter to Groves proposing the idea of a laboratory in the northeast, and then became head of the Initiatory University Group, which worked out the specifics. Pegram had also been one of the incorporating trustees of Associated Universities, Inc., or AUI, the organization that ran Brookhaven and served as a buffer between it and the government. The lecture series had been created for eminent scholars to talk about the relation between science and society. Oppie's upcoming lecture series—he was to give three talks—got a lot of attention in the local press. Expecting thousands of people, more than could fit in any of the lab's auditoriums, Brookhaven set up an area for him in the field next to the old theater. The subject of Oppie's talks was complementarity, and although this couldn't have been of much interest to his audience, he succeeded in his usual way, enchanting them during the first part of his first lecture.

I saw Oppie only occasionally during the next two years. I didn't think he looked well. He had always been too thin, but now he began to look painfully so, and frail. In 1965 there was a meeting of the APS in San Francisco. I stayed with Ed and Elsie McMillan. One evening, they threw an open house for the people attending the meeting. About seven Elsie said there was a phone call for me. When I answered, it was Oppie. He asked if I could join him and Kitty for dinner in San Francisco. I told him it would be difficult to get away, because the party was more or less in my honor. But he was rather insistent. I made my apologies to Elsie, who allowed me to go, and drove over the bridge to meet Oppie and Kitty at Jack's. Meeting him there brought back memories of the 1930s. The food was excellent; I still remember that Oppie told me to order a mutton chop. But the mood seemed a little subdued. When we left, I walked the Oppenheimers to their car. As I was leaning in the window to say goodnight, Oppie told me that he had just heard from the hospital that his throat cancer had recurred.

In the middle of February 1967, on a Thursday, Gian Carlo dropped by my office and told me that he had visited Oppie in Princeton the day before and that Oppie really looked terrible. I resolved to go down the next day to see him, but something came up that made my presence at the physics colloquium desirable, so I postponed my visit until Monday. On Sunday morning, I read in the *New York Times* that Oppie had died on Saturday night.

Kitty called me up a day or two later and said that she was planning to have a memorial service at Princeton for him on February 25, and asked me to speak. Charlotte vetoed the idea. She said the task would be difficult for me; I wasn't good at that kind of speaking and wouldn't do a good job. I more or less agreed. The service was held at the auditorium at the Institute, and I remember that Charlotte and I sat behind Kitty and Toni. The speakers were Hans Bethe, Henry Smyth, and George Kennan. Afterwards, Kitty took Oppie's ashes to St. John and scattered them near Carvel Rock, a rock about a mile or so outside Hawksnest Bay, which she could see from her house.

I overruled Charlotte when, later that spring, I was asked to be one of the speakers at a memorial session for Oppie to be held at the Washington meeting of the APS in April. For this occasion I had a

Fig. 8.6 Charlotte, about 1967.

long time to prepare, and went back and read all of Oppie's papers, giving them some critical analysis. That's the kind of thing I could do well. The talks were published in the October 1967 issue of *Physics Today*, under the title "A Memorial to Oppenheimer"; in 1969 they appeared in a volume entitled simply *Oppenheimer*, published by Scribner's. I sent a reprint of the *Physics Today* article to General Groves, and in his letter of reply, dated July 13, 1967, he wrote, "I was pleased with the emphasis you placed on his ability to work in close collaboration with the experimentalists and his most unusual ability to inspire his students and his sensitive perception of their problems. These outstanding characteristics were the basis for my selection of him for the Los Alamos post which he filled so admirably."

A couple of years before, Charlotte had left her theatrical work and taken a job with Louis Harris, who was just gaining prominence as a pollster. At first, she was just one of his interviewers, but he soon became impressed by her abilities enough to ask her to become manager of his office. But she refused and in fact quit her job. I couldn't understand it. I soon found out that Charlotte had Parkinson's disease. She had had the flu in 1919, and Parkinson's was a not uncommon aftereffect. She consulted Dr. George Cotzias, at Brookhaven, who was working on Parkinson's disease. He gave her advice but was unable to provide a treatment. Charlotte took it very hard. She had always been proud of her dexterity, and now she was unable to control the shaking of her hands. I told her that it really wasn't very noticeable, and that she controlled it well in public. But she was not comforted. Perhaps I could have found some more sympathetic approach. She became reclusive, not wanting to be seen in public. She became depressed—so badly that she had to spend some time in psychiatric treatment at New York Hospital.

On May 23, 1967, Rabi retired from Columbia. The university gave him a dinner in Low Library. There was never a question of us not attending. The night before, she took a bottle of sleeping pills, and in the morning I found her dead. I called Rabi from across the hall, and he summoned an ambulance. Henry Barnett made the funeral arrangements.

N I N E

New York and St. John, 1968–1997

In 1968 Kitty had the idea of commemorating Oppie's death by holding a theoretical physics conference on its anniversary. With the help of friends, I arranged a one-day conference held at the Institute for Advanced Study, modeled on the Shelter Island conferences of two decades before. Attendance was limited to twenty-five people. Most of them arrived the night before and stayed at the Princeton Inn, where I had drinks and an ample spread of food from Zabar's. After the conference, Kitty gave a reception at her house, featuring unlimited quantities of good French champagne and all the fresh caviar one could eat. For a few years this conference continued as an annual event.

In these years I was doing a lot of advisory work. The directors of many of the labs were old friends: Ed McMillan at Berkeley, Maurice Goldhaber at Brookhaven, Wolfgang Panofsky at Stanford, Bob Wilson at Fermilab, and Louis Rosen at the Los Alamos linac. I was a consultant to Fermilab, the new accelerator laboratory outside of Chicago, before the staff moved out to the site and was still housed in a building in Overbrook, not far from O'Hare airport. I did at Fermilab the same kind of thing I had done at Berkeley after the war: I gave lectures on the state of particle physics. Once a week I would fly from New York to O'Hare, drive to Overbrook, and give my lecture. Afterward, if the weather was nice, we would sometimes lunch on the grass in front of the office building in the European outdoor dining style you would expect of Bob Wilson, with bread, cheese, and good wine. After an afternoon of consulting, I would fly back home. It was strenuous, particularly in winter. There was one morning when we were hit by a cold spell and the temperature at La Guardia Airport was well below zero (it was nineteen below at O'Hare). After we boarded the plane we sat at the gate for twenty minutes with nothing happening. Finally the hostess got on the intercom and said, "We're having an emergency. The coffee is frozen. I'll take a vote of the passengers on whether we should depart without hot coffee." I continued my weekly trips after Fermilab moved out to its proper site, until the time when they got a theory group of their own.

I was also an adviser for the Stanford Linear Accelerator Center while its first big linac was being built. Part of the committee's charge was to see that SLAC was a genuine user's laboratory. We ran into some new difficulties with this one. The experimental measuring apparatus was becoming so large and specialized that it would be difficult for an outside user to work without collaborating with members of the staff. Also, the SLAC staff deserved some recompense for all the hard work they had devoted to these detectors. Later on, from 1978 to 1982, I visited Los Alamos a couple of times a year as a member of the program advisory committee of the Los Alamos Meson Physics Facility. And from 1968 to 1971, I was a member of a committee, chaired by Val Fitch, whose job it was to recommend what new accelerator should be built at Brookhaven. We finally decided on an intersecting storage ring accelerator with superconducting magnets, nicknamed ISABELLE.

The late 1960s was a time of protest on American university campuses. At Columbia the Physics Department was a particular target because a number of the professors were members of Jason. Jason, a division of the Institute for Defense Analysis, was a summer consulting group for national defense, sponsored by the Department of Defense. Charlie Townes, who was vice president for research of the Institute for Defense Analysis, once asked me to become a member of Jason. But I declined, partly as a consequence of the experience I had had with the Navy in connection with that postwar conference in Japan (although Charlie assured me that he would have no difficulty having my security clearance reinstated) and partly because of my repugnance at the Vietnam War—although I thought there was a lot of truth to the argument that if one didn't do the advising, some real SOB would.

One day a group of protesters consisting of students and young faculty, not from Columbia but NYU, took over Pupin lab, the building housing Columbia's physics department. T. D. Lee got involved in negotiating to get them out. It was reputed that he had some good French wine stored in his Pupin office. After some time, he managed to establish that they wanted to be arrested. Since that was also his idea, the two sides were able to come to an agreement. But after the big riots on campus in 1968, the police refused to come on campus, so it was arranged that the protesters would leave the building and cross the sidewalk for their arrest. However, they made one more condition. Before being arrested, they wanted to be taken to Chinese lunch. T. D. obliged, and after lunch they returned to Pupin, crossed the sidewalk, and turned themselves over to the police.

At the beginning of 1969, I was recruited to become vice president elect of the American Physical Society (APS). Nowadays, elections are held between competing candidates, but then it was different. There were no policy issues which were thought to require a choice between candidates, and each year the process followed the same pattern: a nominating committee would pick a candidate for APS vice president elect, the candidate would be informed of the choice, he would refuse (there weren't yet any women presidents), and then the nominating committee would twist the candidate's arm until he agreed. The candidate would serve for a year as vice president elect, serve for another year as vice president, and then automatically become president the following year. At the end of his term, he had two duties. One was to give the retiring president's address at the ceremonial session of the annual meeting in New York, and the other was to be chairman of the nominating committee to select the next year's vice president elect and become chairman of the APS Council.

Nominees were always reluctant to serve, because presidential duties were a diversion from their research. Rabi had been chairman of the nominating committee that had nominated Fermi. When he informed Fermi, Enrico flatly refused. Rab then told him the committee had done its work and adjourned, and if he didn't like the result the only thing he could do was persuade the committee to meet again and choose another candidate. Rab left it at that and Fermi's name went to the electorate.

At the beginning of 1969, John Bardeen was the outgoing president and chairman of the committee that nominated me, but Luis

Alvarez was president of the society, and I think John got Luis to approach me to take the job. Charlotte would have said it was not my cup of tea and told me to refuse. But by this time I was under the influence of Kitty Oppenheimer, who had very different ideas. So in 1969 I became the APS vice president elect and in 1970 vice president. It was my first experience as a science administrator. Ed Purcell, who was president of the society, delegated some not too onerous tasks to me and was very helpful in getting me used to my new and forthcoming responsibilities.

But the political ferment of the 1960s was finally catching up with the APS. In 1968, at the Chicago meeting, while Bardeen was president, anti-Vietnam War activists disrupted a session of invited speakers, to try to prevent presentation of papers by speakers from the weapons labs, Los Alamos and Livermore. In 1969 the protesters formed an organization called Scientists and Engineers for Social and Political Action (SESPA), one of whose purposes was to politicize the APS.

My last duty as vice president was to chair the first half of the ceremonial session at the APS annual meeting, held on February 1, 1971, in New York. The ceremonial session was a joint session of the American Physical Society and the American Society of Physics Teachers. In the first half, which I chaired, Ed Purcell gave the retiring presidential address, and I then relinquished the chair to someone from the physics teachers, who introduced their speaker, Ed Land. Land was the president and director of research at Polaroid, who had been invited to deliver a talk on "The Retinex Theory of Color Vision." Polaroid was then a common target of protests. It sold a picture identification system widely used in making driver's licenses in the United States, but was also being used by the South African government. Activists at a SESPA meeting had circulated a leaflet announcing that they would disrupt his talk. Just before Ed began to speak, a black youth sitting in the center front row got up to launch an attack on Ed and the Polaroid Corporation. Surprisingly, Ed seemed overwhelmed—he became visibly upset, began to tremble and shake, could hardly talk, and I think tears came to his eyes. I later learned that the youth had been a protégé of Land's, and that Land had done everything he could to help his advancement. Land was so shaken that he couldn't speak for some moments. When he finally tried to reply he was no more rational than the protester.

Finally Ed Purcell, a good friend of Land's, went up to the podium to calm him and get him to start his talk.

At the end of that annual meeting, I officially became the APS president. The president's duties were in the direction of policy and ceremony, chairing meetings of the American Physical Society's governing body—the APS Council—and presiding at the dinners associated with the various APS meetings. The real work of arranging the meetings and organizing the program was done by Bill Havens, the secretary of the society, who was always a great help and support to the president.

The protests peaked during my year in office, when I found myself in the unexpected position of representing the establishment. It was a time of much dissatisfaction among APS members and especially younger physicists. The Nixon administration was unpopular not only because of its Vietnam policies but also because it had sharply curtailed support for science. More and more, the annual APS meeting was becoming a forum for protesters who wanted the APS to take up social issues and to promote the economic interests of its members.

I was opposed to both developments, which were against the charter of the APS, which was "the advancement and diffusion of the knowledge of physics." The founders of the APS recognized that other problems were of importance and founded the American Institute of Physics (AIP) specifically to deal with the kinds of issues that concerned the protesters. Their complaints should have been directed to the AIP, not to the American Physical Society. While critics regarded preserving the original mission of the APS as elitist, I viewed it as essential to protecting the integrity of science and preventing its political manipulation. There was another good reason for not moving in the direction of giving the APS the functions of a union. To do so would have meant forfeiting the tax exempt status of the organization. The APS did set up a consulting service for unemployed physicists, but that was as far as we went. But the protests had been gaining strength in the previous few years, and a movement developed to have the society consider not only the welfare of physics but also the welfare of physicists.

I was not completely unsympathetic to some of the concerns. In fact, I got the council to appoint a committee whose purpose was to make recommendations on how we could strengthen the second purpose mentioned in the constitution, science education. Jerrold Zacharias was chairman of that committee. I don't recollect that the committee did very much, but it was a precursor to the foundation of the Forum on Science and Society, which considered social questions related to science and made reports that it hoped would be valuable to the general public and to Congress on scientific aspects of policy questions. I did approve the extension of the American Physical Society's activities in this direction. And I was also not unsympathetic to the plight of younger physicists in the grip of a severe depression in the physics job market due largely to the policies of the Nixon administration. At Columbia, a fellowship program which had supported half a dozen postdocs was abruptly cut to zero.

At the Washington meeting, in April 1971, Ed David, who had recently been appointed science adviser to President Nixon, was to be the after-banquet speaker. On the afternoon of the banquet, Bill Havens informed me that he had received a threat from some activists to disrupt the banquet unless some demands were met, including passing out petitions at the banquet deploring the choice of David as a speaker and adding a protest speaker to the program.

Bill made an appointment for me to meet the activists. Maurice Goldhaber, director of Brookhaven, happened to be around, and I took him along to the meeting. It turned out that the activists were Pierre Noyes, Jay Orear, and Maurice Bazin. I knew Noyes and Orear very well; Pierre had gotten his Ph.D. under my direction in Berkeley in 1950. I let Maurice Goldhaber do most of the talking. I refused to let the protesters bring in their petitions. But Maurice came to an agreement that they would be allowed to have someone speak for five minutes in reply to David's talk.

I accepted this arrangement, which was a case of double-dealing on my part. Knowing Noyes and Orear as I did, I had rapidly become convinced that there was no substance to their threat to disrupt the banquet, that they were bluffing. Still, the Nixon administration was under fire. There had been a huge protest march against the Vietnam War in Washington just the weekend before, and I had a chance to throw another brickbat at Nixon in retaliation for his cut in science support.

At the banquet, at the end of dinner, when I got up to take the microphone, a young man in jeans dashed in from behind the speaker's table and grabbed the microphone. He was forcibly removed by a couple of guards supplied by the hotel before having a chance to say anything. When I introduced David, whose talk was entitled, "Is Physics Obsolete?" I made some remarks criticizing the administration's actions towards science and informed the audience that there would be another speaker at the end of David's talk, who had requested five minutes in which to reply to him.

David's talk painted a rather dismal picture of the future for young physicists. Unfortunately, Noyes did not respond to David's talk but began by telling David that unless he resigned his position as science adviser, he was liable to be tried as a war criminal. I cut him off promptly at the end of his five minutes. In retrospect, I feel I was wrong in my double-dealing because it was impolite to Ed, our guest.

More politics followed at the business meeting. The business meeting was an annual formality, and hardly anybody went to listen to such items as the treasurer's report, so it was easy for SESPA to pack the meeting. Steven Neuman introduced a resolution that the APS "publicly disavows any professional support for William Schockley's publication of his racist theories." I pointed out that none of the statements of Shockley that Neuman objected to were printed in APS publications. I also reminded the audience that the object of the society was to "the advancement and diffusion of the knowledge of physics," and ruled the resolution out of order. Somebody quoted the rules of order to the effect that the assembled body could override the chair. I didn't know anything about the rules of order, but Havens assured me they were right. The matter was put to a vote, I was overruled. I then reminded the meeting that, according to the APS charter, only the APS Council could make policy, and that only recommendations could be made at the business meeting. Neuman agreed to reword his resolution as a recommendation to the council, and in this form it was passed. But of course the press paid no attention to this fine point.

Subsequently, when the ballots for vice president elect and new council members were sent out, the protesters presented a petition with enough signatures to get a proposal on the ballot to change the American Physical Society's constitution, but it was defeated by a large majority.

I didn't do very well at the banquet at a later meeting, a meeting of the APS solid state division in Cleveland a little later in the year.

Before the banquet, I was in my hotel room trying to figure out something to say that would keep the audience awake. One part of my job at the banquet was to award the American Physical Society's annual prizes for distinguished work. I noticed something about the recipients of the prizes. When I was starting to speak about it, I pointed out that, although the recipients came from all kinds of colleges, they all got their doctoral degrees at the most prestigious institutions. Maurice Goldhaber, for instance, got his degree from the Cavendish Laboratory in Cambridge. I suggested that perhaps the moral of this was that at a time of very stringent financial difficulty what money was available should be funneled to a few outstanding laboratories to maintain their excellence.

This woke them up all right. Of course, it offended a large majority of the audience, and Bill Havens received a number of letters of complaint later. I should have been more cautious since it was a spur-of-the-moment notion. Even if it were true, which is questionable, I shouldn't have said it.

In 1971, politics was even influencing prestigious awards such as the Oppenheimer Prize. The Oppenheimer Prize was awarded each year by the Center for Theoretical Physics at the University of Miami at Coral Gables, which was run by a scientific council under the direction of Behram Kursunoglu. Kursunoglu had appointed me to a committee, chaired by Marvin Goldberger, to select the 1972 winner of the prize. The choice was easy: Steven Weinberg. Weinberg had been a codeveloper of the electroweak theory, an astonishing development which had suddenly brought field theory back into fashion (and for which he later received the 1979 Nobel Prize).

But when we conveyed our nomination to the APS Council, Weinberg was vigorously opposed by council member Edward Teller. Teller was angry because, not long before, Steve had taken a public stance against the ABM, the anti-ballistic missile system proposed by President Nixon, and Teller's influence over Kursunoglu was sufficiently strong that he succeeded in derailing Weinberg's candidacy. For a while there was a standoff. A compromise was finally worked out (suggested, I'm told, by council member Julie Schwinger), according to which I would receive the 1972 prize, with Steve to win it the following year. When they offered me the prize, I had mixed feelings, and many people, especially Viki Weisskopf, urged me to refuse as a protest against Teller's action. The person who convinced me to take it was Kitty, who felt that a public scandal over the award would dim the luster of the Oppenheimer Prize.

On February 1, 1972, came the annual APS meeting at which I was to give a presidential address and then step down as president. My presidential address was entitled, "Serber Says: Volume III." Most of it was about the state of physics at the time. At the end, however, I made a few remarks about the conflicts of the society during the previous year, and said that I felt that it was wrong to try to change the character of the society. I said it should continue to focus on promoting science, and stay away from looking after the economic interests of the members as well as from taking stands on social issues unrelated to science. After I sat down, one activist came over and asked me to request that the chairman—incoming APS president Philip Morse—give them time for a rebuttal. I refused. This was a ceremonial session. It was a fixed program—not a debate. I think the "equal time" notion got established with the American public as a result of "equal time" rules for television stations broadcasting on controversial subjects. That foisted the idea that all questions had two sides.

That reminds me of something that happened at Columbia in the 1960s and early 1970s. When Columbia's nuclear engineering department had built an experimental nuclear reactor and was preparing to turn it on, some antinuclear factions aroused the local population, sparking a lengthy controversy. They began speaking of the reactor, which at most would have developed a few kilowatts of power, as if it were a thousand-megawatt nuclear generating plant. The academic senate appointed a committee to pass on the question of whether the reactor should be allowed to operate. After hearing evidence, its conclusion was that there was no danger at all from the operation, but in view of the opposition of the local population, the plant should not be activated. A short time after this result was published, a student who had been on the committee wrote a letter to the *Columbia Spectator* saying that while he had voted in favor of the report, he was unhappy about it. The trouble, he said, was that all the experts had testified the same way!

When the nominating committee met, I had a bright idea about who the next candidate should be. I suggested to the committee that it choose Chien Shiung Wu, a Columbia professor and the world's most prominent woman physicist. The committee unanimously approved the idea. When I called Chien Shiung to tell her of her nom-

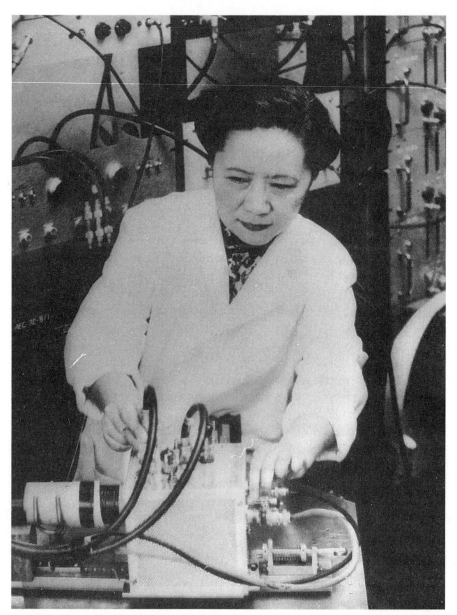

Fig. 9.1 Chien Shiung Wu (1912–1997) was born in Kiangsu Province, China. She came to the United States in 1936 to study with E. O. Lawrence and received her degree from the University of California, Berkeley, in 1940. She joined Columbia University in 1944 and became a specialist in beta decay. In 1956–57, along with collaborators, she carried out an experiment providing the first evidence for parity violation in the weak interaction. She became president of the American Physical Society in 1976.

ination, she of course refused. But I was able to persuade her, by pointing out that there never had been a woman president of the society, and that she owed it to the feminist movement, which she warmly supported, to accept the post.

Feminist concerns were in fact growing in the profession. Fay Ajzenberg-Selove had organized a panel discussion on women in science at the 1971 January meeting of the APS, and afterwards Vera Kistiakowsky got twenty distinguished women physicists to sign a letter to the APS Council to set up a committee. The council was thoroughly sympathetic, and a Committee on Women in Physics was established, with Vera as the first chair. Vera was the daughter of George Kistiakowsky, a chemist who had played an important role at Los Alamos. She had been a research associate at Columbia in 1954–1957, and an instructor in 1957–1959. Vera suggested her committee should rewrite the constitution of the society to remove any sexist language. The only quarrel I had with feminism was with the use of unmelodious words like *chairperson*. So I moved an amendment to her resolution to the effect that the committee should not use slang. My amendment was taken as a joke, and never voted on.

That year, the Oppenheimer memorial conference had a much different flavor than in previous years. As one can imagine, limiting the conference to twenty-five people created tremendous social pressure, which I was able to resist successfully for a few years. However, the pressure on people at the Institute for Advanced Study became so great that they caved in and refused to hold the event unless it was more open. They appealed to Kitty's democratic ideas and sympathy for the younger physicists, and in deference to her wishes I allowed the graduate students and postdocs at the Institute and the university to attend. But this ruined the character of the meeting. With twenty-five people, one can have a genuine back-and-forth discussion on topics; with a much larger audience, that becomes a matter of lecturing, with at most a question-and-answer period at the end. Later, T. D. Lee told Kitty that the purpose of the conference, to discuss the current state of theoretical physics, had been lost. She turned to me and said, "You shouldn't have listened to me."

At the end of the spring semester of 1972, I received the honorary degree of Doctor of Science from Lehigh University, my alma mater. I also began a year-long sabbatical, and Kitty and I had planned to spend most of it on a sailing expedition. In the spring following the

Fig. 9.2   Kitty and I, 1972.

first Oppenheimer memorial conference in 1968, at Kitty's urging, I
had bought a forty-two-foot Rhodes yawl from the publisher of *Time*
magazine, James R. Shepley. During the years I had been sailing on
Long Island Sound, I had become friendly with Phil Gauss, the manager of Minneford's Boat Yard on City Island, where the America's
Cup racers had been built. Out of friendship, he surveyed the boat
for me. Under the name *Andromeda* she had been sailed a couple of
years earlier in the Bermuda race, where she had finished respectably
in the middle of the pack. I planned to sail her to St. John, where Kitty
continued to spend her vacations at the Oppenheimer house on
Hawksnest Bay (which today is a community center). I took the
Andromeda to Minneford's to be fitted out for the trip. One day we
threw a big christening party, attended by most of the Columbia
physics department, and many Brookhaven people, as well as some
other friends. Toni renamed the *Andromeda* the *Undique*, which I did-
n't think was an improvement, but Kitty did; the name derived from
Toni's days as a Latin student. Toni christened her so enthusiastically
that shards of champagne bottle were spread all over the deck.

Since I had no offshore sailing experience, I hired a crew for the
trip to the Virgin Islands from a company whose business it was to
supply yachts with crews. The night before we were to leave, three

Fig. 9.3   Toni Oppenheimer christening the *Undique*.

hefty young college boys appeared; the fourth crew member was John Cool, the son of Rod Cool, one of the senior Brookhaven physicists. Their captain turned out to be quite extraordinary. He arrived at dawn in a sports car, driven by a beautiful young English girl. He kissed her good-bye, unloaded his duffle bag, and came aboard. He was an Englishman of about thirty, by profession a ship's officer. He had just returned, two weeks before, from an expedition of which he was the leader, and which had attempted to be the first to reach the North Pole by dogsled in wintertime. The expedition had failed because the plane that was supposed to drop them supplies couldn't reach them because of bad weather.

We sailed off down the East River through New York Harbor into the Atlantic. The trip was instructive and enjoyable: the warm waters of the Gulf Stream after the chilly start, dolphins swimming under our bow, whales surfacing a few feet abeam. There were only a couple of untoward incidents. Some distance northwest of Bermuda, we saw a red rocket one night and had a chance to observe the operation of one of the traditions of the sea. The captain immediately headed us towards the spot and got on the radio inquiring about the vessel in distress. We found nothing from where he judged the rocket to have been fired, but the captain refused to leave. We criss-crossed the ocean, searching the area and broadcasting queries repeatedly for a couple of hours, until we finally got a response from the radio-telephone. A voice informed us that we were surrounded by ships of the U.S. Navy, blacked out for some naval exercise, and demanded that we get out of the area as fast as we could. The next day, a switch shorted out and discharged our batteries, so neither our running lights nor our engine worked. Fortunately, we weren't too far from Bermuda and could make an unplanned stop there for repairs. After Bermuda we had to traverse the doldrums of the Sargasso Sea. In the old days, sailing ships without engines must have had a hard time of it. We had to stop occasionally to clear the seaweed from our propellers. We finally arrived at Kitty's triumphant but damp. As usual in such circumstances, there wasn't a dry bit of bedding or clothing aboard.

During the next three years, we made several trips in the *Undique* down the Leeward Islands, as far as Granada. On one of these I had as a guest Tulio Regge, a physics professor at the Institute for Advanced Study, and his wife Rosanna Cester, who was also a physicist. On a small island in the Grenadines, Tulio sat under what must have been a manchineel tree (though I didn't recognize it), because he got some of the sap on his fingers and rubbed his eyes, and for the next twenty-four hours was blind and in considerable pain. A number of Columbus's sailors were said to have died from eating the "death-apples" of that tree. In 1972 I sold the *Undique* to a friend on St. John, Bob Eaton; when I visited St. John in August 1995, he was still sailing her. He had changed the name to *Galatea*.

During my sabbatical in 1972, we planned a trans-Pacific trip to Japan. We planned to cruise for a while in the Caribbean, pass through the Panama Canal, and head for Japan via the Galapagos Islands and Tahiti. Kitty's daughter Toni and her husband had bought a boat in Sweden and were leisurely sailing around the world. When my sabbatical began, they were crossing the Pacific, and Kitty hoped to catch up to Toni somewhere in the eastern Pacific. For the trip, Kitty bought a new boat, a fifty-two-foot ketch which was named *Moonraker*. The boat had been built in Hong Kong a couple of years before, completely of teak, had elegant interior fittings, and fine wicker work that would have cost a fortune in the States. She had a 160-horsepower Diesel engine and carried enough fuel to carry her a couple of thousand miles. Her name had two meanings:

Fig. 9.4   Moonraker, in Hawksnest Bay, St. John, August 1972.

someone touched with madness, and the topmost sail on a full-rigged ship. She was berthed in Ft. Lauderdale at the Bahia Mar Marina, made famous as place where the fictional detective Travis McGee kept his houseboat.

In May she was sailed, with a crew picked up in Ft. Lauderdale, to San Juan. As is usual in such cases, I was stalled there for a couple of weeks waiting for a single-sideband radiotelephone to be delivered and installed. From there with some friends to St. Thomas where I was stalled another two weeks waiting for some new sails and spars to be finished. Kitty had recruited a crew of unexpected origin: four young men from New Hampshire whose professional trade was carpentry. They had come down to get jobs and enjoy life in the Virgin Islands, but she convinced them that what they really wanted to do

Fig. 9.5 Kitty
Oppenheimer, 1972.

was sail across the Pacific. Then we embarked on a route that took us
to Martinique, Granada, Bonair, and on to South America.

We approached the coast of Colombia just at dawn one morning.
Everything was covered with a heavy mist. As the sun just began to
rise, in the distance we saw a mountainside. The fog rose and rose,
higher and higher, and there was more and more mountainside, to an
incredible height. Finally, there were the peaks at perhaps a thirty-
degree angle. I was astonished; I had no idea that there were mountains over 22,000 feet high just beyond the Colombia coast.

We went west to Cartagena, Colombia, and from there visited the
San Blas islands on the way to Panama. We entered the San Blas
islands from the eastern end, where tourists seldom go. The only
airstrip was one hundred miles further west; the eastern end was
only visited by an occasional private yacht. The Indians still lived a
very primitive life. Typically, they had to row a couple of miles to the
mainland to get water. We traded trinkets, brought for the purpose,
for *molas*, the embroidered panels decorating the women's shirts. I

bought a very nice one for twenty-five cents. We toured the San Blas for several days, but then Kitty began to feel ill. She continued to grow weaker, and it suddenly became clear that she was really sick. We cut off the rest of our cruising and headed straight for Panama.

We reached Cristobal, at the Atlantic end of the Panama Canal, on October 17, and had to take an hour-long train ride across the isthmus to Panama City, where she was admitted to Gorgas Hospital. There she was put to bed immediately and was told that she had a severe intestinal infection. She never left the hospital and lived another ten days. On October 27, she died of an embolism. During this time, I slept on the boat and took the train back and forth each day to be with her. Once, during these miserable days, I was mugged after coming out of a grocery store in Colón while walking down the main street in the middle of the afternoon. An arm reached around my neck from behind while two hands reached into my rear pants pockets and deprived me of my wallet and traveler's checks. They dropped the traveler's checks, and when I bent down to pick them up, the two men ran away and dashed down a side street. There was only about five dollars in cash in the wallet, but I regretted losing my New Mexico driver's licence that said "No Name Required."

Toni had left a probable itinerary, and I had cabled her shortly after Kitty entered the hospital. She flew to Panama immediately, but arrived only the day after Kitty had died.

Everyone we had dealings with in the Canal Zone was very kind to us. The governor of the Canal Zone had us in to see him to offer his condolences and to make sure we were getting all the help and comfort possible. Toni and I turned the *Moonraker* around and started the return trip to St. John with Kitty's ashes. The way back was against the trade winds and we made slow progress. After a few days we put in at Les Cayes, a port on the south shore of Haiti. We dropped anchor in the harbor and put up a quarantine flag to signal customs ashore, and in a short time a customs outboard came out to inspect us. There was a customs officer, a policeman, and a young man in his early twenties. Toni, who was an excellent linguist and had worked as a three-language interpreter at the UN, understood their lingua franca. They went through the usual formalities, and at the end the customs officer and policeman got in the outboard and left, much to our surprise leaving the young man aboard. On cross-examination, he disclosed that he was a police spy and would stick with us as long as we were in Haiti. This was our introduction to the dictatorship of Duvalier. Actually, we found him quite useful. He ran our errands, bought food for us, got our laundry done ashore, and in general made himself useful while reporting all these developments to police headquarters. We were appalled by the poverty we saw in Les Cayes, which was almost as bad as what I had seen in India.

After about three or four days, we left again for St. John. When we arrived, we scattered Kitty's ashes off Carvel Rock, where Kitty had scattered Oppie's five years earlier.

Toni and I left *Moonraker* on St. John in care of her crew and flew up to Princeton to straighten out legal matters with Governor Minor, the Oppenheimers' attorney. There we met Peter, Kitty's son and Toni's brother, whom Toni had phoned. We returned to St. John in December, and at the end of that month sailed for Florida. On the way we ran into a huge storm. The wind was not all that high—it averaged about thirty-five knots, with gusts never over forty-five— but the storm was steady, lasting for three days. The waves built up to a height of thirty feet, and the constant stress wrought consider- able damage to *Moonraker*'s sails and rigging, and we were in less than shipshape condition when we reached Ft. Lauderdale and the marina, the same spot we had left eight months before. The *Moonraker*'s crew spent a few weeks refurbishing her, and in short order she was sold. That was the last yacht I captained. My yachting career ran a common course; as it progressed the boats got larger and larger, until they reached the point where one couldn't afford them any longer.

Toni didn't return to her husband, but got a divorce and moved into the house at St. John; a year or two later she remarried. Kitty had built for my use a small guest house—actually, just a room and bath—close to the main house, and I continued to spend all my free time on St. John.

When my sabbatical was over, I returned to my teaching duties at Columbia. I haven't said very much about my teaching, but of course over the years much of my time was devoted to teaching graduate students in physics, to directing the research of graduate students, and to serving on departmental committees. At Columbia at one time or another I taught quantum mechanics, nuclear physics, elec- tricity and magnetism, and group theory. I served for a while as chairman of the Graduate Committee, where I introduced one

Fig. 9.6 Toni Oppenheimer.

notable reform. Graduate students working for a Ph.D. were required to take a qualifying exam near the end of their first year, a written examination lasting several days to test their likelihood of successfully completing the Ph.D. program. If they failed, they were allowed to repeat it twice more in succeeding years. Of course, the students were terrified of the exam, and many of them postponed it until their second year of graduate work. They came up with all kinds of excuses—I heard about any number of dying grandparents. I adamantly refused to accept any excuses and forced all the students to take the exam their first year. Most of them passed and were thereafter grateful to me because I had saved them a year in their work for the degree. There was also the Fellowship Committee, which was a wearing job; one had to read more than a hundred applications. Applicants were required to write a short thesis telling why they were interested in a career in physics. The most notable reply came from a student in Korea, who began his, "After sweethearts, I like physics best." The committee was strongly tempted to admit him, but unfortunately his grades were too low to permit it.

There was also the Ph.D. final exam committee, the last hurdle a candidate had to pass. We sometimes found that our attempts to implant a broad education in physics didn't survive a couple of years of intensive specialized research. One student in particle physics from the Nevis Laboratory was asked how he would measure the angle of a prism. This was a standard optics lab experiment; a telescope was used with the prism mounted on a graduated turntable that could read the angle to a minute of arc. The candidate replied, "With a protractor, I suppose." Another candidate was asked to state the second law of thermodynamics, and replied, "I don't know them by number."

I had been teaching a long time and I often met people who reminded me that they had taken a course from me. This led me to invent a fable: I died and went to heaven, and Saint Peter led me into the presence of God, who said: "You won't remember me, but I took your Quantum Mechanics course in Berkeley in 1946." A real instance occurred in 1972, when the United States first reestablished relations with the People's Republic of China. T. D. played a large role in the academic aspects of the new ties between America and China, and the Chinese Academy sent over a high-level scientific delegation that visited Columbia. T. D.'s office is right next to mine, and one day when I arrived I bumped into T. D. leading the delegation to his office. He introduced me to one of the senior physicists of the delegation, who spoke almost the same words.

In 1975 I was dragooned by the other members of the Physics Department to become its chairman, a job I held until I retired. It sounded like a formidable responsibility. The budget of the Physics Department was about $6 million a year, and it had about four hundred employees. However, I found that most of the work could be delegated. Leon Lederman was director of the Nevis lab and took a large part of the responsibility. Will Happer ran the Columbia Radiation Lab, Chien Shiung Wu ran nuclear physics, and about all I had to do for them was sign applications for grants which they prepared and occasionally represent them before the government agencies they were applying to. Most of the departmental work was taken care of by Henry Foley, who was the department's college representative, by Gary Feinberg, chairman of the Graduate Committee, and by other committee chairmen. Then there were very capable administrative assistants: Ann Bolton, the long-time department secretary; Guy Castle, the department's business manager; and Irene Tramm, who ran the theoretical section of the department. The only things I couldn't get out of doing were conferring with George Frankel, the Dean of Arts and Sciences, on such things as salaries and new appointments. At the time, the university was going through a period of severe financial crisis, and what one could do about salaries was very limited—mainly controlling the distribution of what funds were available. As far as new appointments went, the Physics Department was about the only one in the university able to offer new appointments. We weren't so fortunate, though, in getting our offers accepted. Promising physicists with young families often weren't attracted to life in New York City, preferring places like Princeton or Stanford.

With the dean's help, I think I did a reasonable job of the day-to-day running of the department. Pupin Laboratory badly needed to be refurbished, but due to the financial strictures of the time, that had to be postponed. But, ultimately, I don't think I made a very good chairman of the department; I didn't plan for its future, didn't succeed in making the appointments that were needed.

One day I received a memorandum from Grayson Kirk, president of Columbia, addressed to all department heads. It said that a new prize had been established in Israel, which carried an award of $100,000, for scientific work with humanitarian or social aspects, and asked for possible Columbia nominations. I thought of Chien Shiung Wu, who had been using nuclear techniques to study sickle-cell anemia, and submitted her nomination to President Kirk. He forwarded it as Columbia's choice, and she was awarded the first Wolf Prize in 1978. And earlier, in 1975, Jim Rainwater had won the Nobel Prize, sharing it with Aage Bohr and Ben Mottelson, for work in nuclear physics.

While I was chairman I published one paper. A paper by Lee and Wick had used a mathematical technique which suggested to me a nuclear model that illustrated some of the principal features of nuclear structure. It was not supposed to be very realistic, but it was simple enough to be easily understood by students. I submitted the paper to the *American Journal of Physics*, but the editors thought it was too difficult for their readers and suggested I submit it to the *Physical Review* instead, which published it in 1976 under the title, "A Simple Nuclear Model" (52).

While serving as chairman I had another university position. In 1974 I was appointed a trustee of Associated Universities, Inc. It had been formed by a consortium of nine universities and was governed by a body of trustees, two from each university, one a scientist and the other a business manager. In addition to managing Brookhaven, AUI also managed the National Radio Astronomy Observatory in West Virginia.

While I was a trustee we had a couple of successes. One was construction of the VLA, the Very Large Array radio astronomy installation in New Mexico. Another was the construction of the National Synchrotron Light Source (NSLS) at Brookhaven, which provided intense beams of light and X-rays for experimental work for physics, biology, and medical investigations. But we also had a serious problem in the ISABELLE project. This ran into unexpected trouble. Two full-scale models of superconducting magnets had been built, both of which operated satisfactorily up to the specified 50 kilogauss magnetic field, and Brookhaven went ahead and let industrial contracts for mass construction of magnets. However, when the first magnets came in, they didn't meet the specifications. It wasn't clear why, so Brookhaven tried building some more magnets on its own, and didn't succeed in duplicating the good performance of the first two magnets. That they behaved so well appeared to be a fluke. The project was in trouble, and floundered along.

Columbia had a strict retirement policy in those days, and I was faced with mandatory retirement when I turned sixty-eight. Towards the end of the 1978 academic year, in April, Chien Shiung Wu put together a retirement party for me. I was very pleased that my old teacher, John Van Vleck, was able to attend; he had just won the 1977 Nobel Prize in physics. Van and Rabi were the after-dinner speakers at the Faculty Club. There was a symposium the next day entitled "Frontiers of Fundamental Physics." The next year, in March 1979, I was invited to a party given by Harvard's Physics Department in honor of Edwin C. Kemble, who was Van's professor, in celebration of his birthday. Kemble, who was ninety, was Van's professor; Van, his student, was eighty, while I, Van's student, was seventy. Van's birthday was March 13, mine March 14, and Kemble's was January 28.

Even though I had retired, the university's president, William J. McGill, asked me to continue on as Columbia's AUI trustee. Things continued to go badly with ISABELLE, especially after 1979. Brookhaven's director at the time was solid state physicist George

Vineyard, and while both he and project director James Sanford were able physicists who would have been effective administrators in easier times, they weren't the kind to take the drastic action that was needed to save the project; we needed somebody like Bob Wilson. Vineyard and Sanford kept trying to improve the magnet without a major change in design. At the AUI trustee meetings, the scientific trustees complained loudly and often bitterly. At least once, a few of the scientific trustees, led by George Snow, tried to get the board to change directors. But the half of the AUI trustees who were administrative trustees representing the business end of AUI tended to support Gerald Tape, the head of AUI, who had worked as a Brookhaven administrator ever since 1950. Tape supported Vineyard, who, after all, had the NSLS to his credit.

When Robert Wilson came to Columbia, I wrote the president that he should be appointed AUI trustee in my stead, and I retired as trustee early in 1981. That July, Nick Samios was put in charge of the ISABELLE project, and eventually named lab director, but it was too late. By then, the physics community had become divided about the project, which was fatal. Some, led by Leon Lederman, began to argue that if ISABELLE were canceled, high-energy physicists could get something better—specifically, what turned into the Superconducting Supercollider (SSC). I was against that argument, on the "a bird in the hand" theory. It turned out that I was almost wrong. The Department of Energy went ahead and canceled ISABELLE in 1983—and Congress canceled the SSC a decade later.

In January 1977, at her home on St. John, Toni Oppenheimer committed suicide in a fit of deep depression after her second divorce. The previous year I had become friendly with Fiona St. Clair, the daughter of old-time residents of the island whom I knew well. Fiona owned a store that sold batik dresses of her own design. In 1978 Fiona moved to New York to work as a freelance fabric designer, and we were married in 1979. She had a son, Zachariah, by a previous marriage, who was four years old at the time. In November of 1980, our son William was born. Fiona gave birth in Columbia Presbyterian Hospital, and it so happened that Bob Wilson was there at the same time for a bypass operation. His room was on the floor below hers, and for a while there was a large procession of physicists visiting Bob on the sixth floor and then coming upstairs to say hello

Fig. 9.7   Fiona, Will, Zachariah, and I at the fortieth anniversary celebration of the Los Alamos lab.

to Fiona and peek at baby William on the seventh. Zach became a favorite of Rabi, still our neighbor across the hall, and Rab once gave us a backhanded compliment in saying, "I wonder where Zach got his courtly manners?"

In the fall of 1983, a petition was signed by about ten thousand scientists worldwide asking for the end of nuclear bomb testing and the production of nuclear weapons. The idea was that it would be presented to the heads of state of each country around the world. President Reagan refused to accept it. Jim Cronin, Phil Anderson, and I presented it to the Secretary-General of the UN. A report of our meeting appeared the next day in the *New York Times* and on TV.

Just before Thanksgiving that year, a made-for-TV movie called *The Day After* was broadcast on national television and in many places abroad. The movie was a simulation of what happened to the city of Lawrence, Kansas, and its inhabitants after it was hit by an atomic bomb blast at the beginning of a war. At that time Zach was in fourth grade at Collegiate School. His teachers, who had heard that I had been involved with the Manhattan Project and had visited Hiroshima and Nagasaki shortly after the bombs had been dropped on those cities to study the damage, asked me to speak at an assembly which would discuss the film. I got them to also invite Henry Barnett to give medical information. We spoke to a couple of hun-

Fig. 9.8 Working at the Oppenheimer house on St. John, with Zach.

dred kids in the school's auditorium. Our talks seemed to elicit surprise and the question period afterward gave us the impression that much of the audience had been brought up to believe dropping the bombs was a sin. They came from a society whose notion of war was formed by Korea and Vietnam and remembered nothing of the realities of an all-out war like World War II.

In 1983 Los Alamos celebrated its fortieth anniversary. I was the chairman of the opening session at which Rabi was the principal speaker. During the ensuing years, I published a couple of articles discussing developments in physics that I had witnessed (53, 56). In the mid-1970s, to pass the time while I was at St. John, I began writing up some of the material from my nuclear physics course and worked on it off and on until 1987, when T. D. Lee found a publisher; the book was entitled *Serber Says: About Nuclear Physics* (54). In 1992 I published an annotated version of my *Los Alamos Primer* report, in collaboration with Richard Rhodes, whom I had met at Rabi's a couple of years before (55). In 1993 I gave the keynote address at the first (classified) session of the fiftieth anniversary celebration of the Los Alamos laboratory. There were armed guards at the doors, and Zach and I were the only ones admitted without a security clearance.

Fig. 9.9   Teaching Zach to walk.

When Zach had to go to the bathroom, he was escorted by a guard. At one point we met Edward Teller, who was then at Stanford. I introduced Zach, saying that he was at Columbia, and, as usual, Ed sought to upstage me. "I have *two* grandchildren at Stanford, so I have that over you," he replied. Zach and I were led out as soon as my talk was finished and we were not allowed to hear the rest of the session. In 1994 I was invited to give the Pegram lectures at Brookhaven; the present book grew out of those talks.

The Collegiate School, which Zach attended, prides itself on being older than Harvard. When he was in the second grade, he was diagnosed as being dyslexic. Fiona elected to tutor him herself, and did so successfully, which is highly unusual for a mother and her own child. In the process, she became interested in the problem of learning disabilities. Now her interests were turned in a new direction. She decided to go back to school, and received a B.A. degree from

Fig. 9.10 Zach, Will, and I.

Skidmore College and in two more years an M.S. and M.E.D. from Bank Street College of Education. Until the spring of 1996, she worked part-time at Collegiate School, as the learning therapist for the upper school and teacher of a fifth grade English class. In addition, she has a private practice tutoring learning-disabled adolescents.

After graduating from Collegiate, Zach went to Columbia College, where he majored in biophysics and held jobs in the genetics laboratory, as a resident adviser who counsels other students in his dormitory, and as a teaching assistant in biology and genetics. In the summers of 1995 and 1996 he worked in the Department of Biology at Brookhaven with a group at the National Synchrotron Light Source. Will attended Collegiate for three years, then the Cathedral School on the grounds of the Cathedral of St. John the Divine. While there he sang in the Cathedral's choir for a couple of years. Coming from an irreligious home he found much to surprise him. One day he asked his mother, "Why do they sing so much about sheep?" For his seventh, eighth, and ninth grades he switched to Fieldston School, the Riverdale campus of the School for Ethical Culture, which Oppie had attended.

In 1996 Zach graduated from Columbia College. In celebration of the occasion, I joined the academic procession at Columbia's commencement that May 15. It was only the second time that I had attended a Columbia commencement. The first was when I escorted Murray Gell-Mann to receive an honorary degree from Columbia in 1977. The day after Zach's graduation, he and I flew to Madison, Wisconsin, where, on May 17, I received an honorary Doctor of Science degree from the University of Wisconsin, sixty-two years after my professor, John Van Vleck, had insisted I attend the Wisconsin commencement so he could escort me to receive my Ph.D.

In the fall of 1996, Zach won a fellowship from the St. Andrew's Society, awarded to a student of Scottish descent, to study at the University of Edinburgh. Will saw an opportunity for adventure and decided to live with his brother and attend a school there, the Edinburgh Academy. Fiona and I visited Edinburgh in November 1996 and found it a charming town. The boys have a flat in New Town, which dates to the early nineteenth century. The physicist James Clerk Maxwell was born in a house two doors from their flat. He also attended Will's school. The furnishings of the school look unchanged; Will may well be sitting at the same desk that Maxwell used. I am pleased that Will is interested in science. He is taking courses in physics, chemistry, math, computer science, and English.

I once heard Willy Fowler say that he didn't know why they paid him for doing what he did, because he would have done it anyway. I felt the same kind of enthusiasm; I would have done physics as long as I had enough to live on. Looking back, I am amazed at the changes that have taken place in science since I entered the field. The one thing that has not changed is the enthusiasm I see in Zach and his friends for their subjects, interdisciplinary subjects that did not even exist when I was a graduate student.

Notes

Introduction

1. See, for instance, Margaret Brenman-Gibson, *Clifford Odets: American Playwright: The Years from 1906 to 1940* (New York: Atheneum, 1981); Madelin Leof's quote is on p. 124); Eric A. Gordon, *Mark the Music: The Life and Work of Marc Blitzstein* (New York: St. Martin's, 1989).

2. Laurie M. Brown, Abraham Pais, Sir Brian Pippard, eds., *Twentieth-Century Physics*, 3 vols., jointly published by the Institute of Physics Publishing, London, and the American Institute of Physics Press, New York, 1995; see esp. chapters 5 ("Nuclear Forces, Mesons, and Isospin Symmetry") and 9 ("Elementary Particle Physics in the Second Half of the Twentieth Century").

3. Laurie M. Brown and Helmut Rechenberg, *The Origin of the Concept of Nuclear Forces* (London: Institute of Physics Publishing, 1996); Lillian Hoddeson, Paul W. Henriksen, Roger A. Meade, and Catherine L. Westfall, *Critical Assembly: A Technical History of Los Alamos During the Oppenheimer Years, 1943–1945* (Cambridge: Cambridge University Press, 1993).

4. For the story of quantum electrodynamics and its renormalization, see Silvan S. Schweber, *QED and the Men Who Made It: Dyson, Feynman, Schwinger, and Tomonaga* (Princeton: Princeton University Press, 1994). For a nontechnical account, see Robert P. Crease and Charles C. Mann, *The Second Creation: Makers of the Revolution in 20th-Century Physics* (New York: Rutgers University Press, 1996), chapters 6–8.

5. See D. Cassidy, "Cosmic Ray Showers, High Energy Physics, and Quantum Field Theories: Programmatic Interactions in the 1930's," *Historical Studies in the Physical Sciences* 12: 1–40.

6. For more on the history of isotopic spin, see Laurie M. Brown,

"Remarks on the History of Isospin," in K. Winter, ed., *Festi-Val—Festschrift for Val Telegdi* (North-Holland: Elsevier, 1988).

7. A. Pais and O. Piccioni, "Note on the Decay and Absorption of the θ°," *Physical Review* 100:1487–89 (1955); K. Landé, L. M. Lederman, and W. Chinowsky, "Observation of Long-Lived Neutral V Particles," *Physical Review* 103:1901 (1956).

8. Robert Serber, *The Los Alamos Primer: The First Lectures on How to Build an Atomic Bomb*, annotated by Robert Serber, edited with an introduction by Richard Rhodes (Berkeley: University of California Press, 1992).

9. Nuel Pharr Davis, *Lawrence and Oppenheimer* (New York: Simon and Schuster, 1968), 165.

10. Robert Serber, *Serber Says: About Nuclear Physics* (Singapore: World Scientific, 1987).

11. Pais, *Inward Bound* (New York: Oxford University Press, 1988), 367 (remark about Robert Oppenheimer), and Pais, *A Tale of Two Continents* (Princeton: Princeton University Press, 1997), 242 (remark about Kitty Oppenheimer).

12. *The Sciences* (November-December 1995), 3; letters to the editor.

13. Interestingly, in some (consequentialist) ethical perspectives it is argued that the intentions and personal feelings of an agent are not relevant to the ethical character of an action.

14. Spencer Weart, *Nuclear Fear* (Cambridge: Harvard University Press, 1988), 197.

15. *J. Robert Oppenheimer FBI Security File: A Microfilm Project* (Wilmington, Del.: Scholarly Resources, 1978), reel 1, file 7.

1. Philadelphia and Madison, 1909–1934

1. Kusch, P., *Journal of Molecular Spectroscopy* 11:385 (1963).

2. Berkeley and Pasadena, 1934–1938

1. W. Heisenberg, *Zeitschrift für Physik* 77:1 (1932).

2. M. G. White, *Physical Review* 47:573 (1935).

3. J. R. Dunning, G. B. Pegram, G. A. Fink, and D. P. Mitchell, *Physical Review* 47:910 (1935).

4. W. Houston and Y. Hsieh, *Bulletin of the American Physical Society* 8:5 (24 November 1933) and *Physical Review* 45:130 (1934).

5. W. Heisenberg, *Zeitschrift für Physik* 90:209 (1934).

6. L. A. Young, *Physical Review* 47:972 (1935), 48:913 (1935).

7. W. A. Fowler, L. A. Delsasso, and C. C. Lauritsen, *Physical Review* 49:561 (1936).

8. H. A. Bethe, *Reviews of Modern Physics* 9:71 (1937).

9. Abraham Pais, *Inward Bound* (New York: Oxford University Press, 1986), 385.

10. H. A. Bethe, *Physical Review* 50:332 (1936).

11. S. H. Neddermeyer and C. D. Anderson, *Physical Review* 51:884 (1937); J. C. Street and E. C. Stevenson, *Physical Review* 51:1005 (1937).

12. H. Yukawa, *Proceedings of the Physico-Mathematical Society of Japan* 17:48 (1935).

13. H. Euler and W. Heisenberg, *Ergebnisse der exakten Naturwissenschaften* 17:1 (1938).

14. B. Rossi, *Reviews of Modern Physics* 11:296 (1939).

15. A. Proca, *Journal de physique et le radium* 7:347 (1936).

16. H. Yukawa, S. Sakata, and M. Taketani, *Proceedings of the Physico-Mathematical Society of Japan* 20:319 (1938).

17. N. Kemmer, *Nature* 141:116 (1938); *Proceedings of the Royal Society* 106:127 (1938).

18. H. J. Bhabha, *Nature* 141:117 (1938).

19. J. M. B. Kellog, I. I. Rabi, N. F. Ramsey, and J. R. Zacharias, *Physical Review* 56:728 (1939).

20. T. H. Johnson and J. G. Barry, *Physical Review* 56:219 (1939).

21. M. Schein, W. P. Jesse, and E. O. Wollan, *Physical Review* 59:615 (1941).

22. L. D. Landau, *Nature* 141:333 (1938).

23. J. R. Oppenheimer and G. M. Volkoff, *Physical Review* 55:374 (1939).

24. J. R. Oppenheimer and H. Snyder, *Physical Review* 56:459 (1939).

25. F. Bloch and A. Nordsieck, *Physical Review* 52:54 (1937).

26. S. M. Dancoff, *Physical Review* 55:959 (1939).

3. Urbana, 1938–1942

1. R. F. Christy and S. Kusaka, *Physical Review* 59: 405 and 414 (1941).

2. H. C. Corben and J. Schwinger, *Physical Review* 58:953 (1940).

3. W. Heisenberg, *Zeitschrift für Physik* 113:61 (1939).

4. G. Wentzel, *Helvetica Physica Acta* 13:269 (1940); *Helvetica Physica Acta* 13:3 (1941).

5. J. R. Oppenheimer and J. Schwinger, *Physical Review* 60:150 (1941).

4. Berkeley and Los Alamos, 1942–1945

1. Peggy Pond Church wrote a book about Edith Warner entitled *House at Otowi Bridge* (Albuquerque: University of New Mexico Press, 1959).

7. Berkeley, 1946–1951

1. H. R. Crane, *Physical Review* 69:542 (1946).

2. C. Helmholz, E. M. McMillan, and D. C. Sewell, *Physical Review* 72:740 (1947).

3. E. J. Cook, E. M. McMillan, J. M. Peterson, and D. C. Sewell, *Physical Review* 75:7 (1949).

4. J. Hadley, E. L. Kelly, C. Leith, E. Segrè, C. Wiegand and H. York, *Physical Review* 75:351 (1949).

5. R. S. Christian and E. W. Hart, *Physical Review* 77:451 (1950); R. S. Christian and H. P. Noyes, *Physical Review* 79:85 (1950).

6. R. Jastrow, *Physical Review* 79:389 (1950).

7. *Les Particules Élémentaires* (Brussels: Instituts Solvay, 1950), 105.

8. R. Bjorklund, W. E. Crandall, B. J. Moyer, and H. F. York, *Physical Review* 77:213 (1950).

9. J. Steinberger, W. K. H. Panofsky, and J. Steller, *Physical Review* 78:802 (1950).

10. W. K. H. Panofsky, R. L. Aamodt and J. Hadley, *Physical Review* 81:565 (1951).

Bibliography of Works
by Robert Serber

1932

(1) "The Theory of the Faraday effect in molecules." *Physical Review* 41:489
–506.

1933

(2) "The Theory of Depolarization, Optical Anisotropy and the Kerr Effect."
Physical Review 43:1003–10.
(3) "The Calculation of Statistical Averages for Perturbed Systems." *Physical
Review* 43:1011–21.

1934

(4) "Extension of the Dirac Vector Model to Include Several Configurations."
Physical Review 45:461–67.
(5) "The Solution of Problems Involving Permutation Degeneracy." *Journal of
Chemical Physics* 2:697–710.

1935

(6) "The Energies of Hydrocarbon Molecules." *Journal of Chemical Physics*
3:81–86.
(7) "Linear Modifications in the Maxwell Field Equations." *Physical Review*
48:49–54.

1936

(8) "A Note on Positron Theory and Proper Energies." *Physical Review* 49:
545–50.
(9) "Proton-Proton Forces and the Mass Defect Curves." *Physical Review*
50:389–390A.

(10) "The Density of Nuclear Levels." With J. R. Oppenheimer. *Physical Review* 50:391A.

1937

(11) "Disintegration of High Energy Protons." With G. Nordheim, L. W. Nordheim, and J. R. Oppenheimer. *Physical Review* 51:1037–45.

(12) "Note on the Nature of Cosmic-Ray Particles." With J. R. Oppenheimer. *Physical Review* 51:1113L.

(13) "Note on Nuclear Photoeffect at High Energies." With F. Kalckar and J. R. Oppenheimer. *Physical Review* 52:273–78.

(14) "Note on Resonances in Transmutations of Light Nuclei." With F. Kalckar and J. R. Oppenheimer. *Physical Review* 52:279–82.

1938

(15) "On the Dynaton Theory of Nuclear Forces." *Physical Review* 53:211A.

(16) "Theory of Neutron-Deuteron Impacts." With W. Lamb. *Physical Review* 53:215A.

(17) "Note on the Boron Plus Proton Reactions." With J. R. Oppenheimer. *Physical Review* 53:636–38.

(18) "Transition Effects of Cosmic Rays in the Atmosphere." *Physical Review* 54:317–20.

(19) "On the Stability of Stellar Neutron Cores." With J. R. Oppenheimer. *Physical Review* 54:540L.

1939

(20) "Beta-Decay and Mesotron Lifetime." *Physical Review* 56:1065L.

1940

(21) "Production of Soft Secondaries by Mesotrons." With J. R. Oppenheimer and H. Snyder. *Physical Review* 57:75–81.

1941

(22) "Electronic Orbits in the Induction accelerator." With D. W. Kerst. *Physical Review* 60:53–58.

1942

(23) "Nuclear Forces in Strong Coupling Theory." With S. M. Dancoff. *Physical Review* 61:394A.

1943

(24) "Strong Coupling Mesotron Theory of Nuclear Forces." With S. M. Dancoff. *Physical Review* 63:143–61.

1946

(25) "Orbits of Particles in the Racetrack." *Physical Review* 70:434–435L.

1947

(26) "Initial Performance of the 184-inch Cyclotron of the University of California." With W. M. Brobeck. E. O. Lawrence, K. R. MacKenzie, E. M. McMillan, D. C. Sewell, K. M. Simpson, R. L. Thornton. *Physical Review* 71:449–50.
(27) "The Production of High Energy Neutrons by Stripping—I. Energy Distribution." *Physical Review* 72:740A.
(28) "The Production of High Energy Neutrons by Stripping—II. Angular distribution." *Physical Review* 72:748A.
(29) "The Production of High Energy Neutrons by Stripping." *Physical Review* 72:1008–16.
(30) "Nuclear Reactions at High Energies." *Physical Review* 72:1114–15.

1949

(31) "The Scattering of High Energy Neutrons by Nuclei." With S. Fernbach and T. B. Taylor. *Physical Review* 75:1352—55.
(32) "The Spins of the Mesons." *Physical Review* 75:1459A.

1951

(33) "The Capture of π-Mesons in Deuterium." With K. Brueckner and K. Watson. *Physical Review* 81:575–78.
(34) "The Interactions of π-Mesons with Nuclear Matter." With K. Brueckner and K. Watson. *Physical Review* 84:258–65.

1952

(35) "The Effect of Atomic Binding on Nuclear Reaction Energies." With H. S. Snyder. *Physical Review* 87:152–53L.

1955

(36) "Proton-Proton Scattering at High Energies." With W. Rarita. *Physical Review* 99:629A.
(37) "Interaction Between K-Particles and Nucleons." With A. Pais. *Physical Review* 99:1551–55.

1956

(38) "Dispersion of the Neutron Emission in U-235 Fission." With R. P. Feynman and F. De Hoffmann. *Journal of Nuclear Energy* 3:64–69.

1957

(39) "Strong Coupling." With A. Pais. *Physical Review* 105:1636–52.

1959

(40) "A General Transformation of the Symmetrical Pseudoscalar Theory." With A. Pais. *Physical Review* 113:955–58.

1960

(41) "Charged Scalar Strong-Coupling Theory." With H. Nickle. *Physical Review* 119:449–57.
(42) "Limits on Electromagnetic Amplification Due to Complementarity." With C. H. Townes. In C.H. Townes, ed., *Quantum Electronics*, 233–55. New York: Columbia University Press.

1963

(43) "Theory of Scattering with Large Momentum Transfer." *Physical Review Letters* 10:357–60.

1964

(44) "High-Energy Proton-Proton Scattering." *Reviews of Modern Physics* 36:649–55.
(45) "Scaling Law for High-Energy Elastic Scattering." *Physical Review Letters* 13:32–35.

1965

(46) "A Class of Optical Models." *Supplement of the Progress of Theoretical Physics* (Yukawa Commemoration Issue):104–107.
(47) "Shadow Scattering at Large Angles." *Proceedings of the National Academy of Sciences* 54:692–96.
(48) "Response of Superconducting Surfaces to Electromagnetic Waves." Columbia Radiation Laboratory (Special Technical Report).

1966

(49) "High-Energy Scattering." In vol. 1, *Some Recent Advances in the Basic Sciences*, ed. A. Gelbart, 73–89. New York: Academic Press.

1969

(50) "Theory of Atomic Beam Optical Double Resonance Spectroscopy." *Annals of Physics* 54:430–46.
(51) "The Early Years." In *Oppenheimer*, 11–20. New York: Scribner's.

1976

(52) "A Simple Nuclear Model." *Physical Review* C14:718–30.

1983

(53) "Particle Physics in the 1930s: A View from Berkeley." In Laurie M. Brown and Lillian Hoddeson, eds., *The Birth of Particle Physics*, 206–21. Cambridge: Cambridge University Press.

1987

(54) *Serber Says: About Nuclear Physics*. Singapore: World Scientific.

1992

(55) *The Los Alamos Primer*. Edited and with an introduction by R. Rhodes. Berkeley: University of California Press.

1994

(56) "Peaceful Pastimes: 1930–1950." *Annual Review of Nuclear and Particle Science* 44:1–26.

Several oral history interviews with Robert Serber are also located at the Niels Bohr Library at the Center for History of Physics, American Institute of Physics, College Park, Maryland.